Approach and Landing

The McGraw-Hill *CONTROLLING PILOT ERROR* Series

Weather
Terry T. Lankford

Communications
Paul E. Illman

Automation
Vladimir Risukhin

Controlled Flight into Terrain (CFIT/CFTT)
Daryl R. Smith

Training and Instruction
David A. Frazier

Checklists and Compliance
Thomas P. Turner

Maintenance and Mechanics
Larry Reithmaier

Situational Awareness
Paul A. Craig

Fatigue
James C. Miller

Culture, Environment, and CRM
Tony Kern

Runway Incursions
Bill Clarke

Approach and Landing
Tony Kern

CONTROLLING PILOT ERROR

Approach and Landing

Tony Kern

McGraw-Hill

New York Chicago San Francisco Lisbon London Madrid
Mexico City Milan New Delhi San Juan Seoul
Singapore Sydney Toronto

Cataloging-in-Publication Data is on file with the Library of Congress

McGraw-Hill

A Division of The McGraw·Hill Companies

Copyright © 2002 by The McGraw-Hill Companies, Inc. All rights reserved.
Printed in the United States of America. Except as permitted under the United
States Copyright Act of 1976, no part of this publication may be reproduced or
distributed in any form or by any means, or stored in a data base or retrieval
system, without the prior written permission of the publisher.

1 2 3 4 5 6 7 8 9 0   DOC/DOC   0 8 7 6 5 4 3 2

ISBN 0-07-138638-6

*The sponsoring editor for this book was Shelley Ingram Carr, the editing supervisor was Stephen M. Smith, and the production supervisor was Sherri Souffrance.
It was set in Garamond following the TAB3A design by Pat Caruso of McGraw-Hill Professional's Hightstown, N.J., composition unit.*

Printed and bound by R. R. Donnelley & Sons Company.

McGraw-Hill books are available at special quantity discounts to use as premiums and sales promotions, or for use in corporate training programs. For more
information, please write to the Director of Special Sales, McGraw-Hill
Professional, Two Penn Plaza, New York, NY 10121-2298. Or contact your
local bookstore.

 This book is printed on recycled, acid-free paper containing
a minimum of 50% recycled, de-inked fiber.

To Shari, my wife and my best friend

Contents

Series Introduction

The Human Condition

The Roman philosopher Cicero may have been the first to record the much-quoted phrase "to err is human." Since that time, for nearly 2000 years, the malady of human error has played out in triumph and tragedy. It has been the subject of countless doctoral dissertations, books, and, more recently, television documentaries such as "History's Greatest Military Blunders." Aviation is not exempt from this scrutiny, as evidenced by the excellent Learning Channel documentary "Blame the Pilot" or the NOVA special "Why Planes Crash," featuring John Nance. Indeed, error is so prevalent throughout history that our flaws have become associated with our very being, hence the phrase *the human condition*.

The Purpose of This Series

Simply stated, the purpose of the Controlling Pilot Error series is to address the so-called human condition, improve performance in aviation, and, in so doing, save a few lives. It is not our intent to rehash the work of over a millennia of expert and amateur opinions but rather to *apply* some of the more important and insightful theoretical perspectives to the life and death arena of manned flight. To the best of my knowledge, no effort of this magnitude has ever been attempted in aviation, or anywhere else for that matter. What follows is an extraordinary combination of why, what, and how to avoid and control error in aviation.

Because most pilots are practical people at heart— many of whom like to spin a yarn over a cold lager—we will apply this wisdom to the daily flight environment, using a case study approach. The vast majority of the case studies you will read are taken directly from aviators who have made mistakes (or have been victimized by the mistakes of others) and survived to tell about it. Further to their credit, they have reported these events via the anonymous Aviation Safety Reporting System (ASRS), an outstanding program that provides a wealth of extremely useful and *usable* data to those who seek to make the skies a safer place.

A Brief Word about the ASRS

The ASRS was established in 1975 under a Memorandum of Agreement between the Federal Aviation Administration (FAA) and the National Aeronautics and Space Administration (NASA). According to the official ASRS web site, *http://asrs.arc.nasa.gov*

> The ASRS collects, analyzes, and responds to voluntarily submitted aviation safety incident reports in order to lessen the likelihood of aviation accidents. ASRS data are used to:
>
> - Identify deficiencies and discrepancies in the National Aviation System (NAS) so that these can be remedied by appropriate authorities.

- Support policy formulation and planning for, and improvements to, the NAS.

- Strengthen the foundation of aviation human factors safety research. This is particularly important since it is generally conceded *that over two-thirds of all aviation accidents and incidents have their roots in human performance errors* (emphasis added).

Certain types of analyses have already been done to the ASRS data to produce "data sets," or prepackaged groups of reports that have been screened "for the relevance to the topic description" (ASRS web site). These data sets serve as the foundation of our Controlling Pilot Error project. The data come *from* practitioners and are *for* practitioners.

The Great Debate

The title for this series was selected after much discussion and considerable debate. This is because many aviation professionals disagree about what should be done about the problem of pilot error. The debate is basically three sided. On one side are those who say we should seek any and all available means to *eliminate* human error from the cockpit. This effort takes on two forms. The first approach, backed by considerable capitalistic enthusiasm, is to automate human error out of the system. Literally billions of dollars are spent on so-called human-aiding technologies, high-tech systems such as the Ground Proximity Warning System (GPWS) and the Traffic Alert and Collision Avoidance System (TCAS). Although these systems have undoubtedly made the skies safer, some argue that they have made the pilot more complacent and dependent on the automation, creating an entirely new set of pilot errors. Already the automation enthusiasts are seeking robotic answers for this new challenge. Not surprisingly, many pilot trainers see the problem from a slightly different angle.

Another branch on the "eliminate error" side of the debate argues for higher training and education standards, more accountability, and better screening. This group (of which I count myself a member) argues that some industries (but not yet ours) simply don't make serious errors, or at least the errors are so infrequent that they are statistically nonexistent. This group asks, "How many errors should we allow those who handle nuclear weapons or highly dangerous viruses like Ebola or anthrax?" The group cites research on high-reliability organizations (HROs) and believes that aviation needs to be molded into the HRO mentality. (For more on high-reliability organizations, see *Culture, Environment, and CRM* in this series.) As you might expect, many status quo aviators don't warm quickly to these ideas for more education, training, and accountability—and point to their excellent safety records to say such efforts are not needed. They recommend a different approach, one where no one is really at fault.

On the far opposite side of the debate lie those who argue for "blameless cultures" and "error-tolerant systems." This group agrees with Cicero that "to err is human" and advocates "error-management," a concept that prepares pilots to recognize and "trap" error before it can build upon itself into a mishap chain of events. The group feels that training should be focused on primarily error mitigation rather than (or, in some cases, in addition to) error prevention.

Falling somewhere between these two extremes are two less-radical but still opposing ideas. The first approach is designed to prevent a recurring error. It goes something like this: "Pilot X did this or that and it led to a mishap, so don't do what Pilot X did." Regulators are particularly fond of this approach, and they attempt to regulate the last mishap out of future existence. These so-called rules written in blood provide the traditionalist with plenty of training materials and even come with ready-made case studies—the mishap that precipitated the rule.

Opponents to this "last mishap" philosophy argue for a more positive approach, one where we educate and train *toward* a complete set of known and valid competencies (positive behaviors) instead of seeking to eliminate negative behaviors. This group argues that the professional airmanship potential of the vast majority of our aviators is seldom approached—let alone realized. This was the subject of an earlier McGraw-Hill release, *Redefining Airmanship.*[1]

Who's Right? Who's Wrong? Who Cares?

It's not about *who's* right, but rather *what's* right. Taking the philosophy that there is value in all sides of a debate, the Controlling Pilot Error series is the first truly comprehensive approach to pilot error. By taking a unique "before-during-after" approach and using modern-era case studies, 10 authors—each an expert in the subject at hand—methodically attack the problem of pilot error from several angles. First, they focus on error prevention by taking a case study and showing how preemptive education and training, applied to planning and execution, could have avoided the error entirely. Second, the authors apply error management principles to the case study to show how a mistake could have been (or was) mitigated after it was made. Finally, the case study participants are treated to a thorough "debrief," where alternatives are discussed to prevent a reoccurrence of the error. By analyzing the conditions before, during, and after each case study, we hope to combine the best of all areas of the error-prevention debate.

A Word on Authors and Format

Topics and authors for this series were carefully analyzed and hand-picked. As mentioned earlier, the topics were taken from preculled data sets and selected for their relevance by NASA-Ames scientists. The authors were chosen for their interest and expertise in the given topic area. Some are experienced authors and researchers, but, more importantly, *all* are highly experienced in the aviation field about which they are writing. In a word, they are practitioners and have "been there and done that" as it relates to their particular topic.

In many cases, the authors have chosen to expand on the ASRS reports with case studies from a variety of sources, including their own experience. Although Controlling Pilot Error is designed as a comprehensive series, the reader should not expect complete uniformity of format or analytical approach. Each author has brought his own unique style and strengths to bear on the problem at hand. For this reason, each volume in the series can be used as a stand-alone reference or as a part of a complete library of common pilot error materials.

Although there are nearly as many ways to view pilot error as there are to make them, all authors were familiarized with what I personally believe should be the industry standard for the analysis of human error in aviation. The Human Factors Analysis and Classification System (HFACS) builds upon the groundbreaking and seminal work of James Reason to identify and organize human error into distinct and extremely useful subcategories. Scott Shappell and Doug Wiegmann completed the picture of error and error resistance by identifying common fail points in organizations and individuals. The following overview of this outstanding guide[2] to understanding pilot error is adapted from a United States Navy mishap investigation presentation.

Simply writing off aviation mishaps to "aircrew error" is a simplistic, if not naive, approach to mishap causation. After all, it is well established that mishaps cannot be attributed to a single cause, or in most instances, even a single individual. Rather, accidents are the end result of a myriad of latent and active failures, only the last of which are the unsafe acts of the aircrew.

As described by Reason,[3] active failures are the actions or inactions of operators that are believed to cause the accident. Traditionally referred to as "pilot error," they are the last "unsafe acts" committed by aircrew, often with immediate and tragic consequences. For example, forgetting to lower the landing gear before touch down or hotdogging through a box canyon will yield relatively immediate, and potentially grave, consequences.

In contrast, latent failures are errors committed by individuals within the supervisory chain of command that effect the tragic sequence of events characteristic of an accident. For example, it is not difficult to understand how tasking aviators at the expense of quality crew rest can lead to fatigue and ultimately errors (active failures) in the cockpit. Viewed from this perspective then, the unsafe acts of aircrew are the end result of a long chain of causes whose roots

originate in other parts (often the upper eche-
lons) of the organization. The problem is that
these latent failures may lie dormant or undetected for hours, days, weeks, or longer until
one day they bite the unsuspecting aircrew....

What makes [Reason's] "Swiss Cheese" model
particularly useful in any investigation of pilot
error is that it forces investigators to address
latent failures within the causal sequence of
events as well. For instance, latent failures such

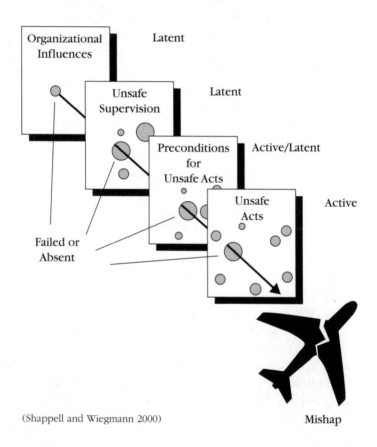

(Shappell and Wiegmann 2000)

as fatigue, complacency, illness, and the loss of situational awareness all effect performance but can be overlooked by investigators with even the best of intentions. These particular latent failures are described within the context of the "Swiss Cheese" model as preconditions for unsafe acts. Likewise, unsafe supervisory practices can promote unsafe conditions within operators and ultimately unsafe acts will occur. Regardless, whenever a mishap does occur, the crew naturally bears a great deal of the responsibility and must be held accountable. However, in many instances, the latent failures at the supervisory level were equally, if not more, responsible for the mishap. In a sense, the crew was set up for failure....

But the "Swiss Cheese" model doesn't stop at the supervisory levels either; the organization itself can impact performance at all levels. For instance, in times of fiscal austerity funding is often cut, and as a result, training and flight time are curtailed. Supervisors are therefore left with tasking "non-proficient" aviators with sometimes-complex missions. Not surprisingly, causal factors such as task saturation and the loss of situational awareness will begin to appear and consequently performance in the cockpit will suffer. As such, causal factors at all levels must be addressed if any mishap investigation and prevention system is going to work.[4]

The HFACS serves as a reference for error interpretation throughout this series, and we gratefully acknowledge the works of Drs. Reason, Shappell, and Wiegmann in this effort.

No Time to Lose

So let us begin a journey together toward greater knowledge, improved awareness, and safer skies. Pick up any volume in this series and begin the process of self-analysis that is required for significant personal or organizational change. The complexity of the aviation environment demands a foundation of solid airmanship and a healthy, positive approach to combating pilot error. We believe this series will help you on this quest.

References

1. Kern, Tony, *Redefining Airmanship,* McGraw-Hill, New York, 1997.

2. Shappell, S. A., and Wiegmann, D. A., *The Human Factors Analysis and Classification System— HFACS,* DOT/FAA/AM-00/7, February 2000.

3. Reason, J. T., *Human Error,* Cambridge University Press, Cambridge, England, 1990.

4. U.S. Navy, *A Human Error Approach to Accident Investigation,* OPNAV 3750.6R, Appendix O, 2000.

Tony Kern

Foreword

Winston Churchill once said, "There comes a time in every man's life when he is called upon to do something very special; something for which he and only he has the capabilities, has the skills, and has the necessary training. What a pity if the moment finds the man unprepared." Mr. Churchill speaks from experience. When his moment came, he was prepared and arguably saved Western civilization. Of course, the average approach and landing doesn't carry quite the same importance as the Battle of Britain and World War II, but shooting an approach down to minimums in crosswinds with the lives of you and your passengers on the line is pretty "special." Most of your approaches will not call for every ounce of knowledge and skill you possess, but if you fly long enough, eventually one approach will quite literally demand everything you have and everything you know. Are you prepared?

For most pilots, the approach and landing is the most demanding part of the flight. Unless you happen to fly acrobatics or in combat, the approach and landing is the "final exam" of the sortie. Whether you fly a single-engine or multi-engine, small or large, for pleasure or pay, getting back safely will require a successful approach and landing. The hazards of weather, winds, poor communication, and apathy are waiting for the unsuspecting and disorganized pilot. With all the challenges to face, some pilots might not know how to improve their performance on the ILS or in the visual pattern.

The aim of this book is to help you meet the various challenges of getting down to the ground, with a methodical approach using extensive case studies, analysis, and plain old-fashioned common sense. The first few chapters outline the scope of the challenge and some larger recommendations dealing with approach planning and stability, and the following chapters take a closer look at specific challenges pilots face during approaches. Tony Kern continues to deliver the clear analysis and plain talk to pilots that he's known for across the globe.

I first met then–Lieutenant Colonel Tony Kern ("sir" to me) when I was teaching aviation to senior cadets at the United States Air Force Academy and Dr. Kern was the head of Military History. Our two departments had decided to create a unique class for cadets based on the airmanship model he discussed in his groundbreaking book *Redefining Airmanship*. I was fortunate to work with Dr. Kern in the design and instruction of the course. He was already an internationally known expert on airmanship and sought-after speaker and trainer. At the time, I was a "good" pilot by my standards, having just left an assignment flying transports in California. Looking back now, it was exposure to the ideas of *Redefining Airmanship* and *Flight Discipline* (another of his books) and, more particularly, exposure to the fire of Tony Kern that actually turned me into a professional pilot. I've never met anybody more passionate about aviation safety and professionalism, and I've been fortunate to contribute to some of Dr. Kern's efforts. His enthusiasm for helping his fellow aviators live long and productive lives is infectious. He is supremely qualified to address this topic and you are in for a treat.

The old flying joke is, "How do you define a good landing? Answer: You can walk away from it. A great landing? Answer: They can use the airplane again." This book will cut through the confusion and show you how to fly better approaches through methodical planning and have "great" landings for the rest of your flying career. It will highlight the challenges you face and give concrete recommendations on how to master them. Like the rest of the Controlling Pilot Error series, this book will help make you a better pilot. So read on, fellow aviator, and I'll see you back on the ground after safe and successful flights for both of us.

Noel D. Fulton

Acknowledgments

I want to gratefully acknowledge the work of the Flight Safety Foundation's Approach and Landing Accident Reduction (ALAR) Task Force, the Federal Aviation Administration, and the Aircraft Owners and Pilots Association for their diligent efforts to reduce pilot errors in the approach and landing phase of flight. It is hoped that this book will contribute to their ground-breaking efforts.

I would also like to thank my friend and colleague Noel Fulton for his efforts on this book, which provided a general aviation and regional airline perspective that I could not. Special thanks (again) go to Shari Kern for her proofreading assistance, and, as always, thanks to the Lord for the gifts He has given me that allow me to contribute to the area of flight safety and professionalism.

Tony Kern

1

The Approach and Landing Challenge

The common wisdom among practicing aviators is that a pilot is only as good as his or her last landing. The point being that performance in aviation is an "all the time" thing, and even among the best, the potential for poor performance lives within us, often very close to the surface. When we get behind the aircraft close to the ground, we are in trouble. Most often we survive through quick recoveries, good "salvage skills," and just plain dumb luck. But with an aircraft fully configured to intentionally fly "slow and dirty," the results of our preparation—or lack thereof—have life and death consequences. That is what this book is about.

Case Study: Just a Routine Full Stop

It was a hot summer day at the southwestern airfield, and the instructor and student were making the last of five practice instrument approaches in their multiengine aircraft. The previous hour of instrument practice procedures had gone without a hitch. The aircraft was functioning perfectly; the skies were clear early but had become more and more cloudy as the day progressed. The prevailing winds were strong and gusty, but basically down the runway. All in all it was a good day for instrument training. The pilots were highly experienced professionals, knew each other well, and were familiar with the airfield and the approach. Just another day—it was all so routine.

As the approach controller gave the crew its turn to intercept final approach, he gave the crew the first hint that this approach might be different than the four previous ones. "Can you accept a short turn to final? I have VFR [visual flight rules] traffic I'm trying to get in on the left." The crew responded with an "affirmative," and the controller quickly followed up, "Turn left heading 210, descend and maintain 3400 until established on the localizer, clear ILS [instrument landing system] runway 18 left...VFR traffic inbound at your five o'clock for the right—he's talking to tower." Both pilots instinctively looked to clear for the potential traffic, neither saw it. It was really marginal weather for VFR traffic, but evidently someone out there was scud running. Hesitant to go "belly up" to the unseen traffic, the pilot in the left seat delayed his turn to the assigned heading momentarily. "Have you got him?" he inquired to the right seat pilot. "Negative" came the reply, "you'd better start your turn." "Coming right" replied the left seat pilot as he rolled the aircraft inbound toward the final approach course. It was clear that they were being turned early and would capture the localizer very near the glide slope intercept point. Scattered to broken low cumulus clouds made a visual approach from this location out of the question.

Simultaneous with the turn, the pilot nosed the aircraft over to descend to 3400 (300 ft) and selected the localizer frequency on his panel instrumentation. In so doing, the pilot neglected to pull enough power and the airspeed increased significantly. The course was already tracking rapidly across and the pilot steepened the turn to capture it. Momentary distraction had resulted in a classic overshoot and the pilot was now likely to stray into the final approach course for the right runway, all the worse because the pilot knew there was VFR traffic inbound to it. Caught in a time-stressed decision mode, the options were obvious, and nearly every instrument-rated pilot has had a similar decision to make. "Do I crank in the turn to capture the course? Do I accept going through the course and start down the glide slope with the hopes of 'recapturing' the course later? Do I embarrass myself and execute a missed approach? Damn that controller! Why did I accept that early turn?" In a matter of seconds, life had gone from routine to downright hazardous, and the lives of two pilots hung in the balance of the next decision—a decision that must be made immediately. We will return to their story in a few minutes.

The "Terminal Approach" Area

The terminal approach and landing area is obviously the most hazardous arena in all of aviation. On one end of the spectrum of operations is a typical large airport, where you have a mix of experimental, general aviation, commuter, and large airline aircraft. The performance and size differences of these aircraft run the gamut from Piper Cub to F-16 to the Boeing 747.

Experience factors range from student pilots on their first unsupervised solo to airline captains and military pilots with 20,000+ h. And of course, you have an air traffic control system attempting to weave all of these aircraft together into a tapestry of safety and efficiency on a few selected and highly saturated frequencies. Oh, and then there is the pilot ego thing, where to ask for assistance or to execute a missed approach in front of the "brethren" loses some serious "cool points."

It doesn't get a whole lot better at the other end of the spectrum, at the small "nontowered" (formerly known as "uncontrolled"—the name change apparently done to remove the connotation that there were no rules to follow) airports. Here, local pilots often develop "local procedures" and deviate from the established "rules of the air" contained in the Aeronautical Information Manual (AIM) and FARs (Federal Aviation Regulations).

To complete the picture of the varied approach and landing environments, we have many areas, such as Alaska, Canada, and some parts of rural America— where airports are an unnecessary commodity, and bush pilots make the vast majority of their takeoffs and landings from wilderness strips or lakes and rivers. In these harsh environments the aircraft is as essential as the family car, and private and commercial aviators must learn extreme pilot techniques just to get around.

The point of this short discussion of landing environments is to illustrate that any comprehensive discussion of approach and landing errors and remedies can present a wide variety of challenges. One thing is clear at this point, however, only through serious personal preparation can a pilot keep the scales tipped in his or her favor. Is it any wonder that this arena results in the most pilot fatalities and is the most unforgiving in all of aviation? But just how bad is it?

The Scope of the Challenge

The focus on approach and landing mishaps and errors is a fairly recent one, even though safety statistics have always illustrated the hazards associated with the high-density traffic areas surrounding airports. Mishap stats also highlight the wide variety of causal factors that result in the relatively disproportionate number of mishaps and fatalities that occur in this flight regime. But although operators, managers, and safety experts have been aware of the challenge of approach and landing mishaps for some time, the serious identification of the types and causes of the errors began in earnest only in the late 1990s, when a group sponsored by the Flight Safety Foundation took on the challenge of identifying the components of approach and landing errors and set out to develop resources and countermeasures to address the problem. Let's look briefly at this monumental effort by the Flight Safety Foundation.

The Approach and Landing Accident Reduction (ALAR) Task Force

The Approach and Landing Accident Reduction Task Force was created in 1996 as an extension of the controlled flight into terrain (CFIT) effort launched in the early 1990s following a series of pilot-caused mishaps on large passenger-carrying aircraft. It is vital to note that nearly all of this important work was conducted by volunteers from within the aviation community, adding to the credibility and, hopefully, the acceptance of the recommendations. As a major component of the study, the group analyzed 621 fatal accidents that occurred between 1980 and 1996. Of this sample, fully 287 of the accidents occurred during the approach and landing phase. From this extensive sample, the group was able to identify a large number of common factors and develop serious recommendations to address them from multiple perspectives, including pilot error, crew coordination, equipment, air traffic control, and approach design. From these recommendations, the task force developed an excellent CD-ROM called the ALAR Toolkit, which contains outstanding training materials and risk management tools. It is available through the Flight Safety Foundation. (*Note:* You can contact the Flight Safety Foundation by mail at 601 Madison Street, Suite 300, Alexandria, VA 22314, or by calling (703) 739-6700. Its excellent website is located at *www.flightsafety.org.*)

Throughout this book, we will refer to the findings from this excellent report, analysis, and recommendations, but there is much more we must do. For all its depth, the ALAR report is somewhat limited in scope in that it addresses only aircraft greater that 12,500 lb take-off weight and specifically excludes smaller piston-driven aircraft, helicopters, and military operators. The largest problem here is that the study excludes most general aviation pilots and aircraft. Although there are certainly transferable lessons for us all from our brothers and sisters in the heavies, a complete look at approach and landing mishaps must be more inclusive and that is our goal with this effort. But we must begin somewhere, so let's take a look at the findings of various studies on the scope of the approach and landing challenge.

Highlights of the ALAR Analysis

The ALAR Task Force reached several conclusions from its study. These are listed to help each of us get a little better handle on recognizing the face of our enemy (it is often the same one we see in the mirror) and provide a platform for action to shore up our personal airmanship and prevent a possible future mishap. All the following statements are taken from the executive summary of the final ALAR report. I have divided the factoids into two categories, the first a general description of circumstances and environmental factors associated with the mishaps and the second, a more focused look at pilot error and interaction.

The General Conditions Associated with Approach and Landing Accidents

There were 287 approach and landing accidents (ALAs) resulting in 7185 fatalities to passengers and crew members. The average ALA rate is 14.8 fatal accidents per year for non-Commonwealth of Independent States (CIS) aircraft. If the trend observed continues, 23 fatal accidents per year can be expected by the year 2010. The ALA rate for freight, ferry, and positioning flights (no passengers carried) is possibly eight times higher than the rate for passenger flights.

Among occurrences where data were available, three-fourths of the accidents happened where a precision approach aid was not available or was not used. Forty percent of the accidents occurred during daylight, thirty-nine percent during night, and two percent during twilight. The accident rate at night is estimated to be close to three times the accident rate during daylight. "Lack of ground aids" was cited in 25 percent of the accidents. The most frequent consequences were collision with terrain followed by postimpact fire.

Other interesting facts from the ALAR study include:

- Nonprecision approaches were primarily associated with CFIT accidents.

- Although 67 percent of all CFIT accidents were in hilly or mountainous terrain, 29 percent—nearly one of every three—were in areas of flat terrain, suggesting that significant terrain features are not a necessary prerequisite for CFIT accidents.

- More than 7 out of 10 CFIT accidents occurred in poor visibility.
- When dual pilot operations were occurring, the captain was the pilot flying (PF) in 74 percent of the mishaps.
- Seventy-three percent of runway excursions occurred on wet runways and in precipitation, and 67 percent involved adverse wind conditions.
- "Incorrect or inadequate ATC [air traffic control] instruction/advice/service" was a causal factor in 33 percent of all occurrences.
- "Disorientation and illusions" was a causal factor in one of every five occurrences.
- "Automation interaction" was a factor in 20 percent of all occurrences.
- "Lack of qualification/training/experience" was a factor in 22 percent of the accidents.
- Formal reports documented multiple cases of "nonstandard phraseology" by both controllers and flight crew members.

Focus on Pilot Error

There were several specific areas identified by the ALA Task Force that are directly related to pilot error, preparation, and performance.

The most frequent causal factor (74 percent) was "poor professional judgment/airmanship" (i.e., decision making).

- "Omission of action or inappropriate action" by a flight crew member was identified in two principal forms and was the second most frequent causal factor.

- Inadvertent standard operating procedure (SOP) deviation (72 percent).

- "Deliberate nonadherence to procedures" (40 percent). This usually referred to a continuing descent below the decision height (DH) or minimum descent altitude (MDA) by the crew without adequate visual reference.

The second most common primary causal factor was "lack of positional awareness in the air, generally resulting in controlled flight into terrain (CFIT)."

Inadequate crew resource management (CRM), such as failure to cross-check or coordinate actions, was the third most frequent casual factor and was present in nearly half the total accidents.

The fourth most frequent causal factor was "lack of positional awareness."

Poor aircraft handling and poor energy management was causal in 45 percent of the mishaps.

- Thirty-six percent (of the total energy management cases) were characterized as "low and slow."

- Thirty percent (of the total energy management cases) were characterized as "high and fast."

These data closely parallel other studies that have been done on an international scale. Dr. Ratan Khatwa, a senior flight deck engineer for Honeywell Aerospace, conducted a study of 4000 routine airline flights and found that circumstances that contribute to unnecessarily high safety risks were found to be much more prevalent than previously suspected. According to his research, factors that could lead to an approach and landing accident are present on nearly one in every five routine flights. The study noted six key pilot "behavioral markers" that were uncovered:

- Procedural errors and violations
- Aircraft mishandling
- Threats from weather and terrain
- Air traffic control events
- Aircraft path or speed deviation
- Aircraft misconfiguration

Khatwa, working closely with the noted aviation researcher Bob Helmriech, recommended action in four critical areas: risk management, standard operating procedures, approach procedures, and aircrew training. We have seen the seriousness of the problem now in professional aviation. Let's turn our attention for a moment to general aviation.

Approach and Landing Mishaps in General Aviation

There are nearly 200,000 general aviation aircraft registered in the United States, used for personal, business, freight, firefighting, air ambulance, fish spotting, and a host of other activities. As diverse as these flight objectives are, each and every flight shares one common goal—a safe landing. According to the Joseph T. Nall report—the annual AOPA (Aircraft Owners and Pilots Association) Air Safety Foundation review of general aviation aircraft accidents that occur each year—nearly 50 percent of all general aviation accidents occur during the descent/approach and landing phase of operation. In 1999, this resulted in 599 NTSB (National Transportation Safety Board) reportable accidents. It is also noteworthy that, while the takeoff and landing phases of flight combine for only approximately 5 percent of the total general aviation flight time, 54 percent of the accidents occur during these two critical time frames.

According to the 1999 Nall report, "the majority of accident sequences begin during phases of flight that take up relatively little flight time but contain the highest number of critical tasks and the highest task complexity." It seems to follow, that if we can become more efficient in these areas, we can improve our personal margins of safety and help reduce the overall general aviation accident rate.

Virtually every analytical study validates the common wisdom that says human error is causal in more than 7 out of 10 accidents. Sometimes it is a skill-based error, such as mishandling the aircraft in a crosswind. Sometimes it is a mental error, such as trying to do something we are not yet capable of. Sometimes pilots are set up by air traffic control mistakes or unreasonable requests. Most of the time, however, it is a sequence and combination of events, like the one reported in the following ASRS (Aviation Safety Reporting System) callback report.

Case Study: A "Tight 360"

Single-pilot operations can also challenge general aviation (GA) pilots, particularly when the flight occurs at night in instrument meteorological conditions, and the pilot is experiencing subtle physical incapacitation. A GA pilot described an episode of spatial disorientation that occurred while attempting to respond to an ATC instruction.

"[During] ILS approach at night in IMC, allowed the aircraft to reach a 60-degree bank before recovery in attempt to comply with ATC request for a tight 360. Did not complete 360° turn. After recovery from unusual attitude, rejoined localizer to airport, switched to local advisory service…without properly canceling IFR [instrument flight rules] clearance after entering VFR conditions." The pilot listed contributing factors in the continuation to his report:

- Pilot was fatigued after 6 h of flight and attack of shingles.

- Pilot should have refused ATC request for a "tight 360." (I question the wisdom of 360° turns at night during an ILS approach at any time.)

- Recovery was delayed by not being on critical instruments while attempting to get flight director to make the 360 and not lose positional awareness relative to ILS course.

- Recent experience not adequate for 360s at night in IMC.

The reporter had flown only a few hours in the last 90 days before the incident occurred. In hindsight, a safer response would have been to inform ATC, "unable 360."

No matter what type of aircraft you fly, the challenge on each and every flight is to reduce the errors you make, and to prevent the errors that can't be avoided from linking into a mishap chain. In the incident above, the pilot made an initial error, but corrected in time to avoid a fatal mishap. Let's now go back to our original case study and see how the story ends.

Final Approach (Initial Case Study Conclusion)

When we left our struggling aircrew, it had flown itself into a dangerous situation in a matter of just a few seconds. If you recall, our pilots were just practicing a bit of instrument work when a combination of low clouds, a scud-running VFR pilot, and an air traffic control request for an early turn to the localizer final had set them up for trouble. They were overshooting the final approach course on a two-runway environment with known VFR traffic approaching to land on the right runway.

Let's pick up where we left off, with the crew needing to make an immediate decision to abandon the attempted approach (execute missed approach) or take aggressive action to salvage a deteriorating situation. Like so many of us, the crew chose the latter. The following events were reconstructed from radar tapes and witness statements to the mishap investigation team.

Likely because of the distraction and early request for the turn, the pilot had not fully configured the aircraft for landing, and while executing the descent to 3400 ft picked up speed, exacerbating the problem of the course overshoot and simultaneously intercepting the glide slope prior to being configured for landing. Recognizing the mistake, the pilot pulled the power to idle and attempted to simultaneously capture the course and glide slope and configure the aircraft for landing. In less than a minute's time, the aircraft impacted the ground one-fourth of a mile short of the runway, with the throttles still at idle, killing both men aboard.

What happened here? We have two qualified pilots flying in their local area with a perfectly functional aircraft in fairly good weather. How could things go south so quickly? Since there was no cockpit voice recorder installed on this aircraft, we can only guess at the crew interaction, but the setup is pretty clear. Most of us have done similar things. Trying to be helpful to an ATC controller and another pilot, these pilots had gotten in over their heads, gotten behind the aircraft, and then tried to salvage a very dangerous situation. Perhaps they had succeeded with similar situations in the past, but on this day, they did not.

Our "Approach"

Let me begin by stating that when I began to think about this book, I saw my own personal background, composed entirely of military and public-use aircraft experience, as inadequate for the task. So I enlisted the aid of a fellow pilot. Noel Fulton has extensive experience in light GA aircraft as well as regional airline experience. His contributions in several chapters balance my perspective and provide a broader depth of experience for case study analysis.

This book is not about second-guessing pilots who have made bad decisions. We have all made bad decisions. Rather, it is designed to assist each pilot with identification of the types of pilot errors that can and do occur in regular patterns during the approach and landing phase of flight. Additionally, we hope to establish and increase your awareness of environmental factors that can set you up for these errors. For example, in researching aviation accidents over the past 10 years, it has become readily apparent to me that self-induced time compression kills more pilots than it should, and I have become hyperaware of it in my own operations. Although I will still take a short vector on occasion, I view any effort to rush me as a serious "red flag," and unless I have all my other ducks lined up, I simply decline the request as a matter of procedure. We will discuss this time phenomenon in more detail in a later chapter. But this book is about more than case studies and lessons learned.

I chose to immerse you in some statistical data to give you a feel for what can go wrong on any given approach and landing, but also to partially convince you that the aviation world has conclusively identified the enemy. Researchers have done us an immense favor in recent years by bringing the hand of science to assist us with demystifying aviation performance. Wherever possible, I will incorporate these studies into my discussion. Armed with the knowledge of the problem, we can begin to undertake the serious work of systematically fixing it.

Overview of the Book

This book integrates many of the pilot errors that are addressed in other volumes of this series, but the time and attention constraints of the approach and landing environment make the errors far more likely to link into a mishap chain with lethal implications. The objective of the following chapters is one of error management—assisting the reader in identifying and mitigating known risk factors before they result in an error, or failing that, to keep the errors from piling up until the pilot is overwhelmed.

Chapter 1: "The Approach and Landing Challenge" gives an outline of typical pilot errors in the landing environment and provides a statistical backdrop for the rest of the book.

Chapter 2: "Approach Plans: Got One?" looks at the basic building blocks of approach planning and points out common pilot errors and oversights.

Chapter 3: "Stability Is the Key" addresses many of the errors identified in the ALAR study, and points out that the key to successful landings begins as early as the initial approach fix.

Chapter 4: "Communication Keys to a Successful Approach" points out common communication errors that can compound quickly in a time-constrained approach, and makes recommendations to correct them.

Chapter 5: "Performance Considerations" looks at the hard science of bringing an aircraft back to earth, and points out some intricacies many pilots may not be aware of.

Chapter 6: "The Big and Small of It: Operations at Large Airports and Nontowered Fields" shows that in spite of their obvious differences, flying into large and small fields share many commonalities when it comes to pilot error—and the answers to correcting them are similar as well.

Chapter 7: "Midair Collision Avoidance" puts the ultimate nightmare for any pilot in clear perspective, and lets you know statistically where the conflicts are likely to occur, and how you can avoid them.

Chapter 8: "Missed Approach: Routine Action or Your Last Chance" points out that many pilots prefer to attempt to salvage a bad approach than to swallow a little pride and go around. This fallacy is deconstructed and concepts identified to help avoid this trap.

Chapter 9: "Wake Turbulence and Windshear" assists pilots in identifying the potential for wake turbulence and windshear and provides techniques that can save lives in these time-critical environments.

Chapter 10: "Sensory Challenges and Illusions" completes the book by pointing out that most of what we do is based upon our perceptions—some of which are occasionally unreliable. Common landing illusions are identified and recommendations are made as to how to recognize and react to them before it's too late.

So let's get started, and begin where all professional pilots do, with preflight planning.

2

Approach Plans: Got One?*

*This chapter was written by Noel Fulton.

It was a great ad campaign. My favorite had a young obnoxious yuppie type walking down a busy city street. He screams into a cell phone, "You're fired," and laughs just before he's hit by a bus. He wakes up in a beautiful room with soft music playing. Seeing a huge plate of chocolate chip cookies, he grabs one, bites into it, and starts laughing. "Heaven," he says, more surprised than the TV audience as he goes to the nearby fridge stuffed with cartons of cold milk. He grabs a carton; mouth full of cookie goodness, but it's empty. He grabs the next carton and the next, getting increasingly frantic until, realizing there is not a drop of cold milk to be had, the ugly truth hits him. "Wait a minute. Where am I?" The tag line pops up in flaming letters, "got milk?" The punch line implies that if you ain't got milk, you're toast...literally.

Switch the scene. It's been a beautiful flight. The late afternoon sun slanting through the clouds makes it seem like your VFR flight plan has taken you through parts of heaven itself. You and your passenger are enjoying the flight and doing so much talking and sight-seeing that you forget to update the weather at your destination. "No big deal," you think. Weather was fine when you left and the flight hasn't been *that* long. You had planned to be down before sundown but the air was so smooth, the scenery so beautiful, and the company so enjoyable, you naturally extended the flight just a bit. You figure you'll get the wheels on the ground just a little bit after sunset. There will probably even be enough light left to see by. You glance at the fuel gauges to confirm that there is plenty of juice to get you there with a touch to spare.

You look ahead and spot the airport beacon. Green: come on in. White: the weather's fine. You finish your checklists, line up on the runway, and start down. Everything is looking good when you suddenly encounter fog about a hundred feet off the ground. Losing site of the runway, you throw in the power and take it around. As you fly over the field, you look down, spot the field and that damn beacon still flashing its lies. "What happened?" your passenger asks. Mumbling something about temporary weather you scramble for the AWOS (Automated Weather Observing System) frequency. You finally get it up and listen intently for a clue to the foggy surprise. "Fulton County Regional Airport. Automated weather observation, one, one, five, four zulu. Winds, two, five, zero at one zero. Temperature, four, three. Dew point, four three. Altimeter." A cold sweat breaks out as you realize what has happened. "Dew point" is the temperature when the air is 100 percent saturated with moisture, and 100 percent saturated can mean *fog*! While you were enjoying the beautiful flight, the sun went down and the temperature dropped to the dew point producing the low ground fog that obscured your approach. Suddenly, the fuel gauge is a touch too empty as your passenger asks, "Got a plan?" In just a matter of moments, you've gone from smug satisfaction to near panic.

Bottom line, most of the challenges outlined in this book can be avoided or overcome through proper approach planning. Communication and crew coordination? Plan it. Missed approaches? Plan it. Windshear and the landing environment? All together now, *plan it!* If you're not prepared to fly the approach, you're not prepared to fly. If you do not have an approach plan that adequately covers the challenges you are about to face, you could end up feeling a bit toasted yourself.

In this chapter we're going to suggest a standardized way to plan your approaches. Use it, adapt it, or toss it and come up with your own, but use something, have a plan. We'll consider a case study where the pilots overlooked or neglected important considerations in their approach planning as an incentive to start using a standardized approach to planning yourself. Fundamentally, about every case study in the book could be attributed to improper planning, so we'll spend more time on the tool in this chapter than on the study.

A Useful Tool

Chances are that if you are reading this book, you've heard of Tony Kern's airmanship model outlined in his 1997 book *Redefining Airmanship* (McGraw-Hill). Dr. Kern introduced a multidisciplinary model of airmanship that brought new understanding to an old word. It sought to answer the question, "What is airmanship?" with a better answer than, "I don't know, but I know it when I see it." (See FIG. 2-1.) Judging from the positive response of pilots and other professionals around the world, he was largely successful. Dr. Kern says it best:

"This analysis reveals three fundamental principles of expert airmanship: *skill, proficiency, and discipline* to apply them in a safe and efficient manner. Beyond these basic principles, five areas of expertise were identified as common among expert airmen. Expert airmen have a thorough understanding of their aircraft, their team, their environment, the risks or enemy, and themselves. When all of these elements are in place, the superior aviator exercises consistently good judgment and maintains a high state of situational awareness." (*Redefining Airmanship,* p. 21)

Dr. Kern spends the rest of the book showing how pilots and instructors can use the airmanship model to evaluate individual airmanship during debriefings, plan a study program, and analyze accidents and incidents. And although the airmanship model is a great "macro-tool" for the aviator, I'm going to show you how it can be just as effective as a "microtool." What follows is one of the many useful uses of Kern's airmanship model. For

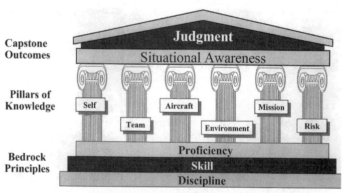

2-1 *The airmanship model—a useful tool for approach planning and execution. Broad-based development—not specialization—is the measuring stick of a professional.*

others, I strongly suggest picking up the book *Redefining Airmanship*.

A Mental Checklist

Long before I'm about to begin my descent and approach, I run through the airmanship model mentally with the approach in mind as a sort of checklist. Maybe you're thinking, "Great! I need another checklist like I need another hole in my head."* But this checklist will work for every aircraft and every organization. The whole point of checklists is to make sure we didn't miss anything important. This model checklist will do the same no matter where, when, or what you are flying. I think about the foundations, each pillar in turn, fly through it mentally to promote situational awareness, and even make some preliminary decisions to help with judgment. Finally, I'll only brief the approach after checking each element of the model. As we go through each of the elements, notice how the airmanship model overlaps in many areas. The foundations and pillars are not separate and distinct. They each overlap and interact with the rest of the model and the rest of the model interacts with each element.

I've been using the model this way for a few years now, so I'm fairly quick with it. It may seem cumbersome at first going through an approach element by element, but with practice, you'll become proficient. I remember when I first started to brief instrument approaches using standard approach plates. It took me 5 min or more but, eventually, I became pretty fluent,

*For more on how proper use of checklists can enhance safety, see *Checklists and Compliance* by Thomas P. Turner, published previously in the Controlling Pilot Error series.

and so will you. Also, on most days, there will be little to analyze. If the weather is "clear and a million" with calm winds, you don't need to spend a lot of time thinking about it unless you are landing behind a larger airplane and wake turbulence is a factor. Think "considered" and move on. The idea is to use a methodical approach that takes the whole picture into account every single time. I think you'll find that walking through the model element by element is a great way to cover everything and thoroughly plan your approach. So, let's get started.

Discipline

I start with *discipline*. Are there any temptations to bend or break some rules? Is it the last leg of a 5-day trip? Am I looking forward to getting home? Why would I feel that I *have* to land here? Do I have a passenger on board that has to be there? Would that passenger pressure me to land, even if it meant extra risk? Would I cave to that pressure? Am I simply tired from a long day and want to get to sleep? Am I so fatigued that I might be tempted to accept less than my best? Do I want to get to the bar before the window closes? Bringing any temptation to compromise flight discipline into *conscious* awareness can go a long way toward keeping your personal commitment to perfect flight discipline. As General George S. Patton said, "There is only one kind of discipline— perfect discipline." Are there any reasons why my discipline on this approach would be less than perfect and are those reasons worth the consequences if something goes wrong?

Skills and proficiency

Next, I go over my *skill* and *proficiency*. When is the last time I landed with these expected crosswinds, this low a ceiling, shot an ILS, or an NDB [nondirectonal (radio) beacon]? How was the last time? Was I high, fast, low, slow, or did everything go well? What kind of weather am I accustomed to on approaches? When was the last time I shot an approach under the forecast conditions? Was it recent? How many have I done recently?

This brings up an important point. Sure, you may be legal with regards to currency, but are you proficient? Different organizations require different levels of currency and the FARs require only six approaches in the last 6 months, but I'm talking about a personal proficiency requirement here. Only you know if you are truly proficient enough to feel comfortable. There have been plenty of times I've been legally current, but far from comfortably proficient. For example, I went to pilot training at Williams AFB near Phoenix. The winds were so calm and predictable there that we'd land runway 30 Monday through Thursday and switch to runway 12 on Fridays. We flew 12 on Fridays on the off chance you had to fly a check ride on the rare day we didn't use 30. Needless to say, I didn't deal with a lot of crosswinds in my student pilot days and neither did the instructors I flew with. The "wing-low" method was academic for me: something I read about but rarely practiced.

Late in the year-long program on one of the frequent cross-country training missions, my instructor and I flew to Vance AFB, Oklahoma, a base known for frequent crosswinds. Because it had been a long flight, we were doing only one approach to a full stop landing. We decided to shoot the VOR approach and I had more difficulty staying on centerline than usual. In hindsight, I realize that I'd intercept the course, line up the needles, and get blown off again. My IP finally told me I could "go visual" and land. "Good," I thought to myself, "I'll put this thing on the ground now." Once I had the runway in sight I noticed a significant crab was keeping me from flying down the centerline, but didn't really do anything else until I was in the flare. I vaguely remember my IP saying something about "not landing in the crab" so I kicked in the rudder to get the nose pointing down to the runway. You guessed it; we started drifting sideways immediately and ended up touching down with quite a jolt because of the side stress on the landing gear. It was hard enough so that we had maintenance check the gear, and luckily, there wasn't any damage. If the wind had been stiffer, there could have been significant damage, maybe even to the point of collapse. There have been other times, long past my student days, when I didn't feel comfortably proficient with certain aspects of an upcoming approach.

The time to realize this and plan for it is long before the initial approach fix (IAF). If you realize you're not comfortably proficient with something you'll face in the very near future, you still have a couple of options. You can use visualization and positive self-talk to chairfly the approach while still at cruise, you can hold to wait if it's a temporary condition, you can go somewhere else, or, if you are flying with another pilot, 'fess up and let the other pilot fly the approach. I saw this happen once. When I was flying C-141s in California, the aircraft commander (AC), recently upgraded to the Starlifter, having spent most of his flying hours in a completely different aircraft, asked another pilot to fly a difficult approach. The other pilot was about to upgrade to AC himself, and *all* of his time was in the C-141. He had about triple the AC's C-141 time. The aircraft commander made the best decision for the safety of the passengers, crew, and airplane. He stepped aside and let another more experienced C-141 pilot shoot the approach. That may be hard and many pilots would rather die than admit they "couldn't hack it," but between you and me, that's just a cliché. I would rather get on the ground safely. The AC above was in complete command, not only of the aircraft and his crew, but also of himself and his ego, and he "hacked it" better by asking for help instead of "John Wayne-ing" it when he didn't feel comfortable. I have to wonder if I could be that mature. I hope I can if the situation warrants it. How about you? How is your approach skill and how proficient are you—really?

Foundation questions (discipline, skill, proficiency) are always tough questions. They are tough to ask and tougher to answer truthfully. But if you cultivate the self-awareness to even ask these tough questions of yourself, you'll have taken a major step toward professionalism. Once you've covered the foundations, you move on to evaluate where you stand on the various areas of knowledge about this approach. I use the acronym STREAM to remember the pillars of the model (self, team, risk, environment, aircraft, and mission). Dr. Kern didn't talk about "mission" in his book but added it later in class discussions and presentations.

For the "self" pillar, we've already done a fair amount of self-analysis asking the tough foundational questions. So this will be pretty quick. Here, I ask how I'm doing. Am I on "high mins" or is this a restricted airport?[*] Am I tired? Do I feel all right? Do I feel irritable or apathetic, possibly because of fatigue? How's my hydration and blood sugar? As little as 3 percent water loss can start to affect cognitive abilities. When's the last time I had some water instead of coffee or a soda? When is the last

---

[*]"High mins" are generally for passenger- and cargo-carrying operators. New captains have to add 100 ft to the minimum ceiling on an approach and $\frac{1}{2}$-mi visibility for the first hundred hours of command in the aircraft. But in no case, may a "high mins" captain fly an approach with weather less than 300-1.

time I had a complex carbohydrate or protein instead of just a candy bar? Hypoglycemia, or low blood sugar, can be an insidious source of fatigue. You just feel tired because your body doesn't know how else to tell you that your blood sugar may be low.* An instrument approach to mins is probably the most demanding thing you'll do during this flight. Are you physically and mentally at your best? Many pilots learn the, "I'M SAFE," acronym in flight training as a good way to evaluate themselves. The letters stand for *illness, medication, stress, alcohol, fatigue,* and *eating.* Sometimes the simple techniques are still the best. If you haven't used, "I'M SAFE," in a while, dust it off. It's worth it.

Next, comes a quick analysis of your team. If you're a single pilot, this will probably go fairly quickly. Just remember that air traffic control (ATC) is part of your team, especially if things go wrong. Remember to use them if it becomes necessary. If there is another pilot on board, it gets trickier, especially if the other pilot is the pilot in command (PIC).

How have your partners been doing? Have they been sharp or have they been missing some of the pilot not flying (PNF) duties? If so, what could be the reason? This may take some tact, but if needed, ask your partners how they are doing and feeling. Ninety-nine times out of a hundred, there will be nothing to worry about. But ask yourself consciously how your partner is doing today on this leg. This is a good time to consider crew communication and CRM. Is there anything special you want from the PNF? Here's a chance to think about how you want to communicate on the approach down to

---

*Richard O. Reinhart, *Fit to Fly,* TAB Books, 1993. If you skip a meal you may not even feel hungry when you have low blood sugar. Hunger is very tied up with circadian rhythms. The rhythms sometimes have more to do with hunger than blood sugar.

rollout and what the respective duties are if something goes wrong. True, a lot of this will be laid out by the organization with which you fly, but be sure to cover any nonstandard items or requests you have as the pilot flying (PF) the approach. There might not be time later.

Once the human element is considered, move on to assessing risk. In a sense, this whole planning process is about uncovering, assessing, and mitigating risk. But here we'll look for some specific types of items. Where are the risks in this approach? Is it the weather? Is it the aircraft? What about windshear and wake turbulence? Is it crew or communication related or something else? Use the risk pillar to identify where the risks are and do what you can NOW to lower them. I usually ask myself, "If I had an accident on this approach, where would it come from?" I've found this question to be extremely valuable in uncovering blind spots and weaknesses, not just in planning an approach, but also in planning flights and in planning my study program to improve my overall airmanship. Try asking yourself the same question. The answer can be very revealing.

Next comes the environment and there's a lot to consider here. In *Redefining Airmanship,* Dr. Kern identifies three separate areas in the overall pillar of environment. They are the physical environment, the regulatory environment, and the organizational environment. It is important to consider each one in turn.

Physical environment

The questions you ask yourself when considering the physical environment should be pretty familiar. Here is where you would consider the forecast weather on the approach. How are the crosswinds, the ceilings, and the visibility? Is the weather good enough for the approach at all? If it's close to being out of limits, start making your alternate plan sooner rather than later. Don't count on the ceiling and visibility staying *just* above minimums. Murphy was a weather forecaster. Should you expect any icing or turbulence and, if so, at what altitudes? Will that alter your desired descent profile? If it has been a long flight, get an update on the weather. Is it better or worse than the forecast when you took off? If the weather is better, congratulations, but if it is worse, why is it worse and how will this affect your plan?

The other major part of the physical environment to consider is the charted approach itself and the runway environment. Review the approach plate and note the courses, altitudes, fixes, frequencies, and so forth. How will you fly it? Where are the most difficult and busy parts of the approach? What do you expect to see when you break out? Where will the missed approach take you? How will you set up your navaids and radios? Where is the high terrain or the highest obstacle in the area? Many pilots will commit the minimum safe altitude and the initial heading and altitude of the missed approach to short-term memory. It's a good idea.

Regulatory environment

Knowledge of the regulatory environment should be a given, but often it's not. If you are not up to speed on the latest air traffic control procedures, it's way past too late when you're flying. So if you feel a bit uncomfortable with your knowledge in this area, the only solution is to spend some quality and quantity study time next time you're on terra firma. Otherwise, a mental review of any pertinent ATC or organizational regulations may be in order here. Most large flying organizations have standard operating procedures (SOPs) for the approach. For example, maybe your organization requires the use of an autopilot during certain types of approaches.

Realistically, this is one of those areas of the model that will probably go pretty quickly. If your knowledge is solid and there are SOPs to follow, think "standard" and press on.

Organizational environment

The final aspect of environment to consider is your organizational environment. If you are flying for an organization (company, squadron, club, etc.) that promotes safety by word and deed, and has a healthy corporate culture, then simply count yourself lucky and move on. If there are problems in your organization, could they manifest on the approach? Is there unofficial coercion to press below mins in bad weather? Is there an emphasis on on-time arrivals that might tempt you to close the gap on a preceding heavy aircraft—putting you in danger of a wake turbulence incident? Are there problems between your union and management? How about within the pilot force itself?

History has shown significant increases in incidents and accidents after organizations merge their operations and their pilot forces. When I was flying C-141s in California, my unit was deactivated and we were merged with the one next door. There were natural rivalries and different attitudes to deal with and everyone was a bit on edge for a little while after the fusion. Considering the organizational environment in this way can give you additional insight into the underlying dynamics of your team. Again, if there's nothing that worries you, press on. But if there is something, bring it out into the open and deal with it in your plan.

The aircraft

Next, consider the *aircraft*. Some aircraft are able to get down and slow down at the same time, some can't. Is there anything about the descent and approach that would make you want to descend earlier or later than usual? Your experience with this aircraft will interact with your knowledge of the weather and of the approach itself. Do you want to delay your descent to minimize time in icing? How will the airplane get down in time for the approach if you do? Is everything working? Flight directors, autopilots, deice or anti-ice equipment, and flaps, and whether or not they are working, can all impact the approach and landing. Take a moment to think about any "gotchas" that might arise from the airplane itself. I don't mean you have to review all the ops limits, memory items, and associated notes, warnings, and cautions, but is there anything about the airplane in general or just specific to this tail number that you will need to factor into your approach plan?

The mission

Consider the *mission* itself. For most of us, this is pretty simple. We want to get from point A to point B safely and that's a good start. Passenger carriers may take a moment here to think about repeat business for the company. Are you on time or early? If, so, maybe I can slow down on the descent to smooth out some of the bumps, if ATC permits. What about if I'm late? When I was flying Beech 1900s with a small regional airline in and out of Denver, we were frequently behind schedule in the summer because of thunderstorms. Because I was responsible for boarding the passengers, I was the first to hear concerns about missing connecting flights. If it were my leg, I'd keep the speed up during the descents to try and make up a few minutes while informing the passengers about the reason for the rougher ride. Looking over my shoulder in the small 1900, I'd get raised thumbs and "thanks" as passengers got off. If you are a single pilot or an instructor, think about how to get the best training on the approach. How can you get the most learning accomplished? What can you add to your plan to best accomplish your goals for the flight and for your own professional development?

Situational awareness

After going through the foundations of discipline, skill, and proficiency and the STREAM pillars, you should have greater situational awareness (SA) about the upcoming approach. In the model, *situational awareness* and *judgment* are outcomes. In other words, if you have the foundation and pillars in place then SA and judgment fall into place. But for the purposes of the model as "microtool," consider SA as a last check. Ask yourself, "Did I miss anything? Do I feel confident about the approach?" If you have worked through the model methodically, the answer will most likely be a solid "yes." But if there is any uneasiness or question in your mind, the SA step is the last chance to really catch it. Use it!

Judgment

Finally, judgment is where you take all of the analysis from your methodical approach to the approach and landing, make decisions, and brief it. If you are a single pilot, go ahead and brief it *out loud* as though your toughest instructor were sitting next to you. You may feel funny at first, but speaking out loud as if you were briefing another pilot will help you organize and express your plan better. Try to anticipate what questions or concerns your imaginary companion would have and try to address them in your brief. If you do this simple step, I bet the plan will be clearer in your mind and much more useful. If you are actually flying with a real person, you'll naturally do the same. After you are done, ask the other pilot what you missed and if he or she has any suggestions. Talk about any concerns until you are both comfortable and crystal clear on what to expect during the approach.

Up to now, we've gone through the model as if you were a single pilot, the pilot in command, or the pilot flying the approach. But, what if you are the second in command and/or it's not your turn to fly this leg? Do you just sit still, shut up, and enjoy the view from the right seat? Of course, the answer here is, "NO!" This is not the place to deal with appropriately assertive fol-lowership in the cockpit. For more on that, check out one of the previous volumes in the Controlling Pilot Error series, *Culture, Environment, and CRM,* also by Tony Kern. But if you want to be the best possible sec-ond in command or PNF, go through the model on your own before the other pilot's brief. This will greatly enhance your ability to participate in the approach plan-ning and the approach itself. It will prepare you with knowledge and questions for the PF's brief. And, if you go through the model every leg and not just the legs you are flying, it'll give you twice the practice using the new "checklist" and you'll get fluent faster.

Again, using the airmanship model in this way may seem unwieldy right now, but give it a chance. I think you'll find that it gets easier with practice and that in practice, it covers everything relevant to the upcoming approach. Maybe you already have a technique you use. If so, maybe you can use some part of the model to enhance your approach planning. Whatever the case, use something. It doesn't seem like the pilots in the fol-lowing case study used anything, and the captain paid the highest price for want of a plan.

Case Study: Just a Short Flight (NTSB Report Number MIA00FA056)

It was just going to be a short flight. A microflight in terms of the 44,000 h of accumulated experience between the two-man flight crew. That works out to 1833 days, or over 5 years of flight time above Mother Earth. The flight was a prepositioning leg from Hilton Head, South Carolina, to Beaufort, South Carolina, to pick up a distinguished passenger. Weather was reported VMC (visual meterorological conditions) and they filed an IFR flight plan. No one expected the tragic results because no one ever expects them. The accident wasn't caused by catastrophic systems failure or freak weather. It seemed to be caused, in large part, by poor planning. After all, it was just going to be a short flight.

The copilot (CP) stated that they departed Hilton Head at about 2022, and activated their IFR flight plan. The PIC and chief pilot of the University of South Carolina athletic department was flying the airplane. They climbed to their assigned altitude, and were given a frequency change to Marine Beaufort approach control. They completed the descent and landing checklist; all flight operations were normal. The flaps were extended to the approach position, the landing gear was lowered, and the propeller was set at 1900 r/min. The airspeed was about 140, decreased to 130, and then down to 120 kn. They were cleared for the approach on about a 6-mi final. The PIC was on the instruments and the CP stated he was outside, and checking the radios. They broke out of the clouds at about 900 ft, and were descending at about 480 ft/min. The ceiling was overcast, ragged, and very dark with no visible horizon. The CP stated he could see some lights, but could not see the airport. He looked back inside the cockpit to check the radios, when he heard a thump. He thought the landing gear had collided with the terrain. The airplane started to cartwheel and eventually came to a complete stop inverted. The copilot survived. The PIC did not.

The crew had accepted and flown an area surveillance radar (ASR) approach. The NTSB found that the pilot-in-command's failure to maintain the appropriate altitude (minimum descent altitude) during an ASR approach resulted in an in-flight collision with swampy terrain. Contributing to the accident was the copilot's failure to maintain a visual lookout during the ASR approach.

Perhaps breaking out of the clouds at 900 ft gave the crew a false sense of security. That sense that everything was all right may have lulled the PIC into continuing his descent below MDA without having the runway firmly in sight. Meanwhile, the CP seemed to be "along for the ride" and not monitoring instruments during the approach. Approximately 13 min elapsed from takeoff to "landing." How could this accident have been prevented?

Obviously, there wasn't enough time to thoroughly plan this approach in the short time it took to fly from Hilton Head to Beaufort. Planning in this case would have had to come before takeoff. What parts of the airmanship model would have been relevant to this case? Before you read on, take a minute to reflect on what the model could have turned up.

- *Discipline.* Don't go below MDA unless you have the airport environment in sight. If planning an approach to an airport at night, consider using an approach with a glide path if it's available. The crew seemed to have skill and proficiency in truckloads. How did they feel? The case study doesn't mention anything here as contributing factors.

- *Team.* Did their experience lull them into complacency? They say that the most dangerous crew is two instructors flying together. Is too much experience a bad thing? Probably not, but it might be worth talking about in a brief. In this case, the CP seemed to rely on the PIC a bit too much. The CP certainly didn't seem to monitor the approach very carefully.

- *Risk.* The main risk in this scenario seems to flow from the darkness and sparse lights at the destination. How could the crew have planned for that?

- *Environment.* The approach into a dark, poorly lit field has already been mentioned. Perhaps the CP was also unwilling to mention any irregularities in preflight or flight to the chief pilot?

- *Aircraft.* The board didn't find anything wrong here.

- *Mission.* Nothing critical here.

It seems that breakdowns in discipline, team, risk, and environment led to poor situational awareness of descent and altitude during the final phase of this ASR approach or a poor decision to continue descent below the MDA.

It's always a bit uncomfortable for most pilots to second-guess their comrades. It's also second nature to think, "I'd never do something like that." Both of those tendencies fit here. We don't know what kind of approach planning they *did* do so it's a bit strange to conclude that they didn't do enough. And obviously, *we'd* never descend below MDA without having the airport in sight. Why then does it happen so often?

Summary

Try using the model during the rest of this book and in your other studies of mishaps. The numerous case studies will provide good practice. Become a more active reader. Instead of just skimming through the case studies, any case studies, and relying on the author to analyze what went wrong, get involved with this book and others like it. Before you read the analysis after the case study, take a moment to run through the model mentally as you read the narrative. Where could the model have helped back in the planning stages of the approach? How could the pilot and crew have avoided the mishap through better planning?

If you are not prepared to fly the approach, you're not prepared to fly. Although it may seem like too much trouble at first, using Tony Kern's airmanship model as a microtool in flight or before you take off can greatly aid your approach planning. By going through each element of the model in a methodical manner, you will cover the major aspects of the approach and uncover the major and minor risks inherent in the approach. It doesn't have to take a long time. It doesn't have to distract you from the cruise portion of the flight. If you don't have your own technique, try using the model. Make a commitment to use the model as I've described for 90 days or the next 100 h. Then evaluate its effectiveness. What have you got to lose except that nagging feeling of missing something important? The next time you face a challenging approach and the other pilot asks if you've "got a plan," you will be able to answer, "Hell yes!"

But even the best plan requires sound execution, and our next chapter talks about the issues and challenges of flying a stabilized approach.

3

Stability Is the Key*

*This chapter was written by Noel Fulton.

Stable. It's a nice, inoffensive-sounding word. "Stable" is a word we don't really mind seeing in our daily lives. "I've got a stable job." "He has a stable girlfriend." "The government and economy are looking pretty stable." "Thank God this bridge is stable!" Generally, people appreciate and strive for a certain amount of stability and avoid uncertainty and chaos. Some would go as far as to say that a certain amount of stability, which may differ from person to person, is the key to living a happy, productive life.

Stability is also vitally important in aviation. Let's go back to some basic aerodynamics and the definitions of stability.* There are two types of stability: static and dynamic. Static stability refers to the *initial* tendencies of a body after being disturbed. Remember the classic visual of a ball and a bowl? If you disturb the ball when the bowl is upright, the ball will initially move

*John D. Anderson, Jr., *Introduction to Flight,* McGraw-Hill, 1989.

back toward the bottom of the bowl demonstrating static stability. If you turn the bowl upside down, balance the ball on top, and then disturb it (say by sneezing), the ball will continue rolling away from neutral, down the bowl, across the table, onto the floor, and out the door demonstrating a statically unstable system. Dynamic stability refers to the system over time. If the system eventually tends to return to neutral after being disturbed, it is said to be dynamically stable. If you disturb a system and it gets progressively more disturbed over time, then you have a dynamically unstable system. And even though it gets complicated, aircraft need a certain stability, either inherent to the design or built in by the engineers, to fly safely. All of this is fine for aerodynamics, but what about stability on our final approaches?

If better approach planning is the key to flying safer approaches, then flying a stable approach is the key to turning your plan into reality. Get stable early, and things get easy. Airspeeds are nailed, glide paths are flown, and landings are smooth. A stable approach gives you time to think and react. A stable approach also gives you the ability to notice important things like gear warnings. Have you ever been so focused on reaching the glide path that you temporarily didn't hear the warning horn or a radio call from ATC? If you are like me, the answer here is probably more often than you care to admit.

If something disturbs a stable approach, it's easy to get back on track, but only if you realize the importance of stability and are consciously aware of the deteriorating situation. If unexpected weather or turbulence upsets the glide path or airspeed, the pilot needs minimum control inputs to get the aircraft back on glide slope or approach speed because the aircraft is already

set up to fly where you want it. It feels good to be configured, on speed, on glide path with the power set. It
does not feel good to be screaming down from above
the glide path or slogging up from below with the wrong
configuration and speed. Stability on approaches, like
life, is something to be desired. And sometimes a stable
approach can mean the difference between life and
death. Whether visual or instrument, stability is the key
to approaches.

Visual Stability and Setting Goals

What is it about clear skies that can lead pilots down the
garden path to disaster? Ever hear about the IQ squat
switch in the landing gear? As soon as the weight comes
off the wheels, pilots and especially student pilots seem
to lose 10 to 15 IQ points.[*] I believe there is a similar
switch involving clear blue skies. This VFR switch can
easily lull some pilots into a false sense of security.
Pilots easily forget basic concepts such as "stability on
approaches" when it's CAVOK.[†] True, there is more of a
margin for error when the sun is shining and the winds

---

[*]When I was teaching aviation and airmanship to seniors at the Air Force
Academy, I used to drill my students in basic 60 to 1 rule applications and
time/speed/distance problems involving simple ratios and multiplication/
division while they were in the classroom. Inevitably, I'd get groans from
the students that they knew their basic grade school math just fine and to
please move on. I'd tell them about the squat switch for a laugh and get
smug looks when I warned them that things like math would be harder
when they were flying. A few lessons later, while in the aviation lab where
students used a commercially available computer-based flight simulator to
accomplish basic flight maneuvers, I used to call out basic math problems
while they were "flying" the sim. If I got an answer at all, it would
frequently be wrong or would take a very long time, demonstrating that the
virtual squat switch was just as effective as the real one.

[†]Ceiling and visibility OK.

are light, but don't let it go to your head. It's still important that you achieve a certain amount of stability on
your approach.

Case Study: Stable Down Under

The "down under" mission out of Travis AFB was a
great trip in the C-141. You would take off from the
California coast in the afternoon and fly to Honolulu.
Thirty-six hours of crew rest would give you an evening
and a whole day to yourself on the beach before taking
off at midnight the next day. You'd keep flying southwest to meet the sun for a breakfast of banana pancakes
while refueling on the American Samoa island of Pago
Pago. You'd take off again and coast in over Sydney
Harbor at Sunset to land at Richmond Field. The
overnight lodging outside Sydney back then was at a
hotel/casino near where the Olympic rowing events
took place. The next day, you'd fly a quick out-and-
back to Alice Springs deep in the outback. Alice Springs
to the Australians is sort of like our Roswell or Area 51;
it's where the Aussie government "secretly" keeps the
aliens and their spacecraft. U.S. Air Force aircraft flying
there regularly just added to the conspiracy speculation.
It was always fun to tell a local you were with the U.S.
government and flying up to Alice Springs. After
another night of crew rest outside Sydney, we'd return
home via Waikiki. Like I said, it was a great trip.

I was the copilot on this trip and it was my leg back
from Alice Springs. We were on the ground long
enough at Alice Springs for me to go downtown and
buy an Akubra outback-style hat that I had under a seat
in the back. It was a weekend so things were pretty
slow at Richmond Field. We were told to expect the
visual and were given a descent at pilot's discretion

down to pattern elevation. We were already lined up on the runway for one of those 50 mi straight-ins, although we couldn't quite make out the field. Things were pretty slow so the pilot challenged me to a little game. He said I could reduce power one time and add power one time to get us on the ground. Many pilots play these sorts of games. There's nothing inherently wrong with them. They are useful for learning your airplane better, for gaining a deeper understanding of descent planning, and for building confidence. Descent games are particularly common. As long as you don't put the aircraft in an unsafe situation to avoid losing, the games can be a lot of fun while improving overall airmanship.

I accepted and decided to delay my descent past where I'd normally start down, figuring it'd be easier to lose altitude than gain it under the circumstances. I planned to add power somewhere down the line to catch a normal glide path. As we closed in, I finally pulled power and started down when I thought I might be in danger of overflying the field. Things were looking fine until I had to raise the nose to slow the C-141 below 250 KCAS before continuing below 10,000 ft. "Uh oh," I thought, "looks like I'm going to be long now. No problem, I'll throw out the flaps."

The flaps certainly helped but slowing down to lower them flattened out my descent so I was still way too high. Down came the gear. I threw everything out and even considered opening the window to stick my free hand out to increase my drag and descent. I lowered the nose to bring the speed right up to flap/gear limit speed. "Darn," I said to the pilot, "I don't think I'm gonna make it. I'm going to do some S-turns or I'll have to take it around."* "S-turns would be cheating," he replied.

---

*OK, I probably didn't say "darn" but this is a family book.

Things started to look better but I still didn't think we'd make it. About that time, the pilot did something that made a lot of sense, and continues to strike me as brilliant airmanship today. He set a simple goal. "Crew, this is the pilot. If we are not stable on a good visual glide path within 10 kn of approach speed by 500 ft AGL, we're going to take it around." The C-141 continued to drop and I hit the beautiful "red over white" on the VASI (visual approach slope indicator) with enough time to slow down and meet the pilot's stated goal. We landed without incident although it was a little long, and I won the challenge proving once and for all that "even a blind squirrel finds a nut once in a while."

That simply stated goal, made by the pilot on final approach, was the key. At some point on the approach, whether visual or instrument, you *have* to hit certain parameters and you *try* to hit a bunch more. On glide path, on speed, crossing height and speed threshold, to say nothing of being properly configured are all goals we have on approaches. Usually, they are unsaid—which is normal. But if you start to get behind the aircraft, you've got to set a proper goal and meet it to get back on track. All of the goal-planning books I've read agree that a proper goal includes time. The goal above was a great example of a SMART goal. It was specific, measurable, action-oriented, realistic, and timely. It's not enough to say, "I need to get on speed and on glide path or I'm going to take it around." It's better to say, "If I'm not on speed and on glide path by the marker, I'm taking it around." Of course, I'm not talking here about being slightly off glide path or slightly off speed. I'm talking about that rare time when you've gotten behind or have been slam dunked by ATC and are diving down to reach the glide path. To this day I talk about goals out loud as I fly my approaches.

I try to state precisely what I want the airplane to be doing. I find that it forces me to pay closer attention and fly more exactly when I've already told the other pilot the speeds, altitudes, and vertical speeds I'm shooting for. This is especially important if they are nonstandard in any way. Try explicitly stating what you intend to do the next time you fly approaches and see how it drives you toward greater precision. It works if you're alone also. I'm a big believer in talking to myself when I'm alone in the airplane. Sure, it sounds odd, but "me myself and I" make a better crew than just "me." When I say what I want to fly out loud, I tend to fly better even when alone. Give it a try. Maybe the following pilot should have tried talking to himself.

Case Study: Scraping the Nose (NTSB Report Number ATL89FA189)

During the descent to the outer marker for a visual approach with an ILS backup, the pilot in command did not descend to a point where a stabilized approach to landing could be completed. Winds were light out of the north and visibility was 4 mi. Although it wasn't "severe clear," the minimums were still high enough for a visual approach. Maybe the descent visibility fooled the crew into thinking that they could get away with a "chop and drop" approach. The airplane reportedly crossed the outer marker with a full-scale fly down deflection on the ILS glide slope indicator. The pilot reported that he attained the glide slope from above when he was about 1 mi from the threshold. Although the NTSB report doesn't make it clear, I wonder if "attained" means that the pilot had the glide slope centered or if it meant that the needle came off the bottom of the case. The airplane was flared for landing at an excessive altitude and touchdown was not made until at least 1500 ft down the runway. The airplane touched down hard, bounced, and on the second touchdown, made runway contact with all three gears or possibly nose first. The nose gear collapsed and the airplane came to a stop on its main gear and the fuselage nose. A bad day.

Perhaps you're thinking that this was obviously an inexperienced pilot. Sorry, the PIC had over 7000 h with a commercial airline transport certificate and ratings out the wazoo. Over 6000 of those hours were multiengine and he had solid recent experience. Personally, it makes me nervous when I read about very experienced pilots making these types of basic errors.

We'd all like to think that the pilots who mess up were somehow defective. Perhaps they were new or not very skilled or proficient. I don't want to say that an airmanship accident can happen to any pilot at any time because I don't believe that. Careful, diligent pilots will seldom make sequential mistakes that lead to an error chain and a mishap. The good habits that carried you safely through your various certificates and training milestones will continue to serve you well. You never outgrow good airmanship. Your experience may not save you if you forget the basic principles of good flying and good judgment. And stability on approaches is one of those things you'll need for the rest of your flying career. This is another case where setting some personal minimum standards and adhering to the stable approach concept would have prevented an accident. Don't let a "cleared for the visual" lull you into throwing stability out the window. If the ILS is there to use then use it. A working glide slope doesn't help when the needle is buried at the bottom of the display because you're too high, or worse yet, because it's turned off.

"It's ATC's Fault" Is a Lame Excuse

Generally, ATC and pilots get along pretty well. Most of the time, the communication between air and ground is accurate and helpful. ATC usually does its level best to help the pilot with special requests and keep the paint of your aircraft on your aircraft, and not on someone else's. Occasionally though, ATC will issue a clearance that is far from conducive to a nice stable approach. Maybe they are trying to squeeze you ahead of another aircraft or open up some space for a takeoff. Maybe they are trying to get you on the ground while Mother Nature is still cooperating. Whatever the reason, it is up to the pilot to decide if he or she can accept clearance for an approach. It is up to the pilot to decide if he or she can do it safely. There is usually plenty of guidance for pilots as to what a stable approach should look like. Controllers don't crash airplanes—pilots do. Failing procedural adherence or guidance, common sense is the last defense. Unfortunately, neither printed definitions nor "horse sense" came into play during the following case study.

Case Study: Tired, Rushed, and Very Long (NTSB Report Number NYC99FA110)

"On May 8, 1999, at 0701:39 Eastern Daylight Time, a Saab 340B sustained substantial damage during landing at John F. Kennedy International Airport (JFK), Jamaica, New York. There were no injuries to 3 crew members and 26 passengers, while 1 passenger sustained a serious injury. Instrument meteorological conditions prevailed for the flight that originated from Baltimore-Washington International Airport (BWI), Baltimore, Maryland. The flightcrew was working a continuous duty overnight schedule, or high speed.* They had reported for duty at 2200 landed at 0130 and awoke at 0445 for a scheduled 0610 departure. Neither crew member reported sleeping at all the previous day and little between the flights. They reported being very tired.

"According to pilot interviews, Air Traffic Control (ATC) data, and the cockpit voice recorder (CVR), the departure from BWI, and cruise flight to JFK was uneventful. The captain completed an approach check-list and briefing, and ATC gave the flightcrew a vector for the instrument landing system (ILS) approach to Runway 4R. Later, ATC advised the flightcrew that the runway visibility range (RVR) was 1600 ft. The controller asked if the flightcrew could proceed with the approach, or if they were going to have to hold until the RVR was at 1800 ft. The captain stated that they needed 1800 ft RVR to initiate the approach. ATC then cleared the flight to turn to 010 degrees and intercept the Runway 4R localizer, and hold southwest of the EBBEE intersection, on the localizer, at 4000 ft.

"The airplane had not reached EBBEE, but was on the localizer course, when the controller informed the crew that the RVR on Runway 4R was good enough for

---

*"High speeds" are not very popular among most crew members. Normally, you report late in the afternoon or evening, fly a trip out to an airport, and get off the airplane but remain on duty. Sometimes, you check into a hotel for 3 to 4 h of "rest" before getting back to the airport for an early morning flight back to the home station all within the maximum crew duty day. Some schedules will even schedule crews for three or four high speeds in a row. They can be very fatiguing because the crew, normally used to daytime fly-ing, radically alters its work-rest cycle with little time to get used to it.

the approach and offered them clearance from their present position, 'or you might be too high. Just let me know....' The captain replied, 'we can take it.' The controller then cleared the crew for the ILS approach to Runway 4R. At that time, the airplane was approximately 4000 ft mean sea level (MSL), and 6.6 mi from the approach end of the runway. The first officer began the approach descent, but the captain extended the landing gear and took control of the airplane. During the subsequent investigation, the first officer stated, 'At that point I fell behind the aircraft. My scan was not 100 percent. I missed the call at the outer marker and the 1000-ft call.' The captain's decision to take the airplane and try for the approach himself effectively rendered him a single pilot from this point onward. The first officer provided little support. It's tough to have crew coordination when one pilot is lost and the other is too busy diving towards the ground to help find his partner.

"Approximately 24 s after issuing the approach clearance, the controller asked the crew, 'You good for the approach from there?' The captain replied, 'We're gonna give it our best.' It seems clear that the controller at least had doubts that the airplane could descend in time. When the investigators asked the first officer if he felt uncomfortable at any time during the approach, he replied, 'No, I relied on the captain to know what he was doing and know where he was.'

"During the descent, the flightcrew received four audible warnings, including one 'sink rate,' and three 'too low terrain' warnings. At 0701:12, the first officer stated 'Okay, there's three hundred feet.' Approximately 7 s later, the captain stated 'Okay, before landing checklist is." The first officer replied 'Three green, flaps zero.' During the approach, the first officer made no other callouts, probably because he was behind the airplane. The flaps remained retracted during the approach. However, after the accident, investigators determined that the captain extended the flaps to 20 degrees, most likely to cover the fact that they had been retracted through the entire approach.

"According to radar data, and the flight data recorder (FDR), the airplane's descent rate reached a maximum vertical velocity of approximately 2950 ft/min while in the weather and on instruments. The airplane crossed the runway threshold [at] about 180 kn. It touched down approximately 7000 ft beyond the approach end of the runway, at 157 kn. The flightcrew applied reverse thrust and maximum braking, but the airplane departed the end of the runway [at] about 75 kn. A Federal Aviation Administration (FAA) Inspector observed approximately 300 ft of skid marks, at the end of the runway. Later, the captain stated that a go-around did not enter his mind at 300 ft *because he was still trying to process information.*" (emphasis added)

If either pilot had been processing the information from their company's flight manuals and procedures they might have remembered that, "(The) stabilized approach concept requires that, before descending below the specified minimum stabilized approach altitude, the airplane should be: In the final landing configuration (gear down and landing flaps)...on the proper flight path and the proper sink rate...and at a stabilized power setting. The minimum recommended stabilized approach altitudes are: VFR—500 FT AFL, IFR—1000 FT AFL." Furthermore, the manuals direct that, "When any approach fails to meet the following stabilized approach criteria, an immediate missed approach (or go-around as appropriate) is *mandatory.*" The crew didn't meet any of these stabilized approach standards.

This crew was lucky to survive this time. The accident chain seems so long and heavy that it's a wonder no one was seriously hurt. Where could the crew have broken the links? If you go back to approach planning and use the model, it seems that thorough planning might have prevented this accident. As far as discipline goes, perhaps the crew could have tried to get some late afternoon rest the previous day. A consideration of "self" and "team" would have revealed significant fatigue with all the associated dangers.[*] Realizing this, the crew could have talked about the effect fatigue has on decision making and judgment and resolved to take its time. The most significant foreseeable risk with this approach seems to be the poor weather at the destination. Normally, if a crew expects an approach down to mins, it strives to be extra stable to give the best chance of breaking out and landing. Environmental considerations seem especially weak. We've already mentioned weather, but company guidelines were ignored along with FAA regulations concerning stable approaches. Maybe it would have been a good idea for the crew to include the company's stabilized approach criteria in its approach brief.

During the actual approach, the best thing that could have been done to avoid this accident was to take another turn in the holding pattern and descend before commencing the approach. I understand the crew's haste. I've been there myself. You want to get the aircraft down while the winds, ceiling, or visibility are cooperating. And it seems like a cheap shot to criticize these crew members while I'm here at ground speed zero, but with all the fatigue and other risks against

[*]See *Fatigue* by James C. Miller, published previously in the Controlling Pilot Error series.

them, rushing an approach seems to be the last thing you'd want to do. One turn around the holding pattern would most likely have prevented this accident. It would have allowed the crew to fly the approach together instead of forcing the captain to act as a single pilot. It would have allowed the crew and the aircraft to stabilize.

Barring the one turn, setting some goals on this approach may not have helped get back on track. Without the turn in holding, this crew was probably too high to ever get stable on this approach. But setting a goal and missing it completely would have been a great way to prompt a missed approach. Sometimes, missing a goal is a very important piece of information. For example, if you have less fuel then you planned at a way point, that is a critical piece of data and you better find out what, exactly, the information is telling you. The same is true for an approach. If you miss your glide path and/or airspeed goal, that could be the final prompt to a missed approach or go-around before it is too late.

The last chance to prevent the damage to aircraft and reputations was a missed approach. And let's face it— that's the last chance to avoid a mishap in a majority of the cases in this book. If things don't feel right, take it around. Pilots will abort a takeoff if something doesn't feel right, sound right, or look right. Why not a landing? If the hairs on the back of your neck are quivering, it might be because you forgot something—like to put the gear down. Every year this occurs scores of times to experienced pilots. It's infinitely better to discover that fact when you reach for the handle on a go-around or missed approach; then there's that split second when you keep descending further than you've ever descended before in the flare.

What is it about a go-around that offends so many pilot egos? Gravity is not that strong, but egos sometimes are. In this case study, the hints to go to missed approach seem overwhelming. Almost 3000 ft/min on the VVI, the numerous GPWS (Ground Proximity Warning System) warnings, the copilot calling out that the flaps were still zero (and of course the copilot should have called for the missed approach), flying past the threshold at 180 kn, and flying past the midpoint of the runway while still in the air were all flashing neon signs to go to missed approach. The captain said he didn't go missed "because he was still trying to process information." It looks like the copilot wasn't the only crew member behind the airplane. When no one is mentally ahead of the airplane, that is a tremendous warning sign, the most severe of red flags on the approach.

One final word about this approach. This seems like a great time to use a little heard but sorely needed word in aviation. Bruce Landsberg, executive director of the AOPA Safety Foundation, advises pilots to, "just say 'unable,'" in his May 2001 column for *AOPA Pilot:*

"A sage bit of advice has been passed along through generations of pilots: Don't bust your posterior, and don't let anyone else bust your posterior for you. The relationship between pilots and air traffic controllers is excellent. We work on the same team to achieve the same objective—to get to the destination safely. However, there are times when the pilot, as the final authority for the safe operation of the aircraft (FAR 91.3), must advise ATC that a particular course of action just isn't going to work and will put a flight in peril. Controllers have the option to refuse a pilot's initial request, but the regulations give the pilot the last word."

Mr. Landsberg outlines plenty of situations where "unable" can save pilots a heap of trouble. True, ATC may not like having to issue a different clearance, but they'd rather go to the trouble of finding a new way for you than have the trouble of notifying search and rescue that there is an airplane missing. If you think you are unable to do something, say so sooner rather than later. It gives you and ATC more time to come up with options. The longer you wait, the less options you have until you run out entirely. And that is definitely not the situation in which you want to find yourself.

Conclusions and Recommendations

Stability, always important on approaches, takes on *critical* importance when weather or other factors raise the risk. If the weather is bad in any way, stability will help you get the aircraft down. If the crew is dysfunctional because of fatigue or other factors, a stable approach will usually mean less workload and greater predictability for the team, let alone for the single pilot. No matter the approach, you need to define and reach a certain minimum level of stability. On a beautiful day, maybe you need only to be fully stabilized ½ to 1 mi out from the field. On days with tricky winds or low ceilings, more stability from the final approach inbound will give you a better chance of putting the aircraft down on the ground. Use the airmanship model to help identify the appropriate level of stability needed for your upcoming approach and talk about it in your plan. The pilot will have to decide and act accordingly.

If you start to get behind on the approach, try setting some immediate goals to help you get back on track. Setting the specific goals with appropriate time limits will force you to fly more precisely and, if you fail to meet them, will help you make a timely decision to take the aircraft around if needed. Missing a goal can actually take some pressure off psychologically and help remove ego from the decision. Sometimes our egos want to push it a bit further. Saying you'll go around unless x and y happens could actually get the ego a bit more on your side when it counts by helping you keep to your stated plan.

Hopefully, this chapter has prompted you to think again about things you probably already know. That's the funny thing about fundamentals. We all know them, but for some reason, we don't practice them consistently. A nice stable approach is a fundamental to good airmanship. And you didn't learn about stable approaches here. You learned about them in primary flight training and you learned about them again in basic instrument training. And maybe you recently shot a less than stable approach and you learned again. At best, I simply reminded you about this important key to flying good approaches. So here's to stability. It's a good thing.

Stability of the crew is also a key ingredient of success in the approach and landing pattern. In our next chapter, we will see how effective communication can help us take advantage of all available information and resources.

4

Communication Keys to a Successful Approach*

*This chapter was written with Noel Fulton.

Flying consistently good approaches under tough conditions places many demands on a pilot and/or flight crew. Central to any successful effort is sound information transfer at the right time. This is just basic communication, but taken to a higher level in aviation with more Draconian penalties for failure.

Let's begin with the basics: talking and listening. Together, they constitute most of our intracockpit and out-of-cockpit communication. Many people have the misconception that just by talking they are communicating and that if they improve their presentation skills or their assertiveness, they will drastically improve their dialogue. Although it is important to send clearer messages, the perfectly constructed "five-step assertiveness statement" will do nothing if no one is listening. Tony Kern outlined the five-step statement in *Culture, Environment, and CRM,* published previously in the Controlling Pilot Error series. It is intended to provide a structure for crew members to effectively and efficiently challenge a potentially unsafe course of action.

"Developing the ability to frankly state concerns is a critical CRM skill. Let's suppose that you are flying with a pilot named Bob, who has decided to cancel his IFR clearance 10 mi out from an airfield in marginal weather and proceed visually for landing. You are uncomfortable because neither of you has ever been to the field and the weather is rapidly changing with fog moving in. Use the following five-step procedure to address the concern.

1. State his name or crew position to get his attention. For example, '*Bob,...*'

2. State your discomfort. 'Bob, *I'm uncomfortable with...*'

3. Frankly state the concern. 'Bob, I'm uncomfortable with *the decision to cancel IFR with the visibility going down.*'

4. Offer alternative. 'Bob, I'm uncomfortable with the decision to cancel IFR with the visibility going down. *I think we should keep our clearance in case we need it.*'

5. Get acknowledgment. 'Bob, I'm uncomfortable with the decision to cancel IFR with the visibility going down. I think we should keep our clearance in case we need it. *What do you think?*'"

Dr. Kern goes on to explain how this advocacy step is part of a larger crew resource management (CRM) loop.

The critical point I'd like to make here is how important the last step of this process is. One of the important functions of the question, "What do you think?" is to test whether or not you have been heard. If there is no answer or there is miscomprehension of your statement, that question will uncover the communication problem and you can retransmit or rephrase your statement.

Then it is your turn to listen. First, you have to listen to find out if your message was received *and* comprehended. Second, you have to listen for what the other person is saying. Most people take for granted that people will listen to what we have to say. You cannot afford that assumption in the air. Remember the old saying, "Assumption is the mother of all screwups." You have to make sure that you are being heard *and* understood by crew members and controllers, and assuring this requires as much good listening as good talking.

Communicating with ATC

Before discussing communication within the cockpit, let's talk about communication with our friends outside the aircraft on the ground. It is vital that pilots improve their interaction with controllers and vice versa. A quick scan of any safety database will reveal numerous instances of mishaps, near mishaps, and accidents directly attributed to poor communication between pilots and controllers. For example, a common breakdown in understanding leads to pilots landing without clearance. But it's not just communication while flying that is a problem. Poor communication can start before the pilot even leaves the ground.

For GA pilots who don't have the benefit of dispatchers and company weather reports, the flight usually begins with a call to the flight service station for a weather report. If things look good, filing the flight plan comes next. This is where communication starts. Before you call, have an idea of the route, destinations, and alternates. Listen closely to the brief. Sometimes, the brief can be a little too detailed, but keep your focus and listen for anything unusual or for anything you don't understand and make sure you ask questions until everything is clear. Once you have an understanding of the current picture, ask yourself how the forecast could change for the worse. Like the old adage says, "You need to hope for the best, but prepare for the worst." Hope, by itself, is not a strategy for success—preparation is.

Case Study: Three Degrees

In *AOPA Online,* Bruce Landsberg outlines a tragic accident in his November 1999 "Safety Pilot" column. A Piper Saratoga crashed while shooting an approach into Memphis, Tennessee. A 3-degree temperature difference from the forecast was a contributing factor to the accident. The pilot took off with a forecast at Memphis of 42°F. Considering the standard lapse rate of 3.5 degrees per 1000 ft, the freezing level was forecast to be around 5000 ft. On arrival, the temperature at Memphis was 39°, effectively dropping the freezing level to about 4000 ft. On arrival ATC cleared the Saratoga to descend to 4000 ft. The pilot requested lower, "as soon as possible." He was cleared lower and was eventually handed off to Memphis tower. He never made it.

Radar data showed that the Saratoga's ground speed gradually dropped off to about 65 kn as the pilot maneuvered for the approach. Eventually, the radar picture showed the Saratoga in a 5000-ft/min descent as it stalled. Witnesses on the ground saw the airplane descend out of a 500-ft overcast in a nose down attitude. The nose pitched up for a moment before the airplane entered a spin and impacted the ground. There were no survivors.

Mr. Landsberg points out that there were numerous PIREPs [pilot (weather) reports] from larger aircraft about icing in the area but that these were not passed on to the Saratoga. ATC's primary job is *separation,* and if they get busy, that is all they will provide you. Never,

ever forget that! I learned that once the hard way when ATC vectored us into a thunderstorm. If there is weather in the area, it is the pilot's responsibility to find out if it's a hazard. Sometimes it may take direct questions of ATC to elicit needed information. It's up to the pilot to decide when these questions are worth the effort and the effort may well be worth your life.

In this situation, a review of the airmanship model before the approach may have highlighted the extra risks associated with icing. The pilot may have been more proactive avoiding or leaving the icing at 4000 ft if he had been more aware of the risks. Adding that extra risk to a knowledge of the absent anti-icing equipment on the Saratoga may have even demanded a diversion. Bruce Landsberg suggests, "When a large aircraft reports moderate ice, that is a mandate for light aircraft to avoid the area." That is good advice.

Standard Phraseology Is the Mark of a Pro

Clear communication with ATC is critically important. The best way for pilots to help the situation is to commit to using standard phraseology in all communication with ATC. It's easy to get lazy, cute, or cool and pilots sometimes develop their own "style" to stand out as individuals. But style won't necessarily translate well to other parts of the country. What may be perfectly clear between a southern pilot and an equally southern controller may be confusing if the southern pilot flies to Maine. Standard phraseology is the answer.

The place to brush up on your phraseology is the back of the Aeronautical Information Manual (AIM). The pilot/controller glossary is *the* source. It's worth an occasional review to make sure you understand the words that both you and ATC should be using. Once you know what the words mean, you have to find the discipline to use them regularly and without fail. This is particularly important if you start flying outside the United States or to other English-speaking countries. When I flew cargo internationally, I quickly found out that foreign controllers' comprehension of English would rarely extend far from standard terms. If you asked for something out of the ordinary, you were likely to get an uncomfortable silence in return. Standard phraseology and clear enunciation won't just help the international pilots; it will help us all, a lesson illustrated in the next case study.

Case Study: Stand By, Please (ASRS Accession Number 466801; Narrative by FO)

"On descent into Laredo, Texas, we were cleared down to 9000 ft. I was the PF and we were in contact with Monterrey approach. The controller started asking us numerous questions about our DME from a certain VOR. We were unable to understand which VOR he meant. The controller was garbled and hard to understand. We thought we received an additional descent to 8000 and kept descending while the controller continued to ask our DME. He then told us we should be at 9000. We climbed immediately and were told that another aircraft had to be vectored away from us. The problem seemed to be the poor English and equipment of the controllers."

Analysis

There seems little the above crew members could have done to help this situation. Perhaps they could have told the controller to stand by with all the DME requests. Once they realized how garbled the radios and how poor the controller's English was, they probably listened harder. In this case, I'd be sure to confirm every clearance. It gets tedious, but in some cases, tedium may be necessary to avoid a violation, or worse yet, a midair collision or controlled flight into terrain.

Finally, let me make one last point about standard communication with controllers. I have a bad habit that is shared by many pilots. I sometimes say "to" and "for" when reading back new altitude clearances. For example, I'd say, "Leaving one-two thousand, climbing *to* two-one thousand or, "Leaving three-one-zero *for* one-five thousand." You can see the difference immediately, but controllers can't tell the difference sometimes between "for" and "four" or "to" and "two" and that can lead to confusion. Leave the "to's" and "for's" out of your transmissions. This simple mistake has killed more than one pilot. Let's now turn our attention to communication inside the cockpit.

Communication in the Cockpit

There have been volumes written about better cockpit communication. Improving communication has been a major goal of the various CRM programs from the beginning, and this is not the place to try to review all of that research. The research has done much to arm the pilot with specific tools like the five-step assertiveness statement and the CRM loop. It has made recommendations to captains and first officers and how to incorporate CRM principles into every aspect of the flight. This research has improved safety and, in my opinion, made the cockpit a better place to work. We'll cover a few of these communication techniques after considering the following case study.

Case Study: Third Time's the Charm (ASRS Accession Number 386084; Narrative by Captain)

"It had been a long day and we were feeling every hour of it. Nonetheless, we continued to work as a team including the FO in the jumpseat. I was the PF flying with a check airman in the other seat. We were flying into an uncontrolled field and had to hold for preceding traffic. ATC's holding instructions were confusing so we asked for clarification. The controller replied that he didn't know what he meant either. After exchanging tired and unbelieving looks in the cockpit, we finally arrived at an understandable clearance with the controller. As the check airman/captain entered the hold into the computer [FMS (flight management system)], we started to pick up icing. As we turned on the icing systems we watched the computer start to turn the wrong direction in the hold. I disengaged the FMS from the autopilot and flew the hold using the heading control. While we were all asking, 'What is it doing?' and concentrating on the FMS, ATC cleared us for the approach. The check airman immediately assumed control of the airplane, stating he rarely got to fly anymore.

"At this point, ATC had left us high on the feeder route inbound to the initial approach point, we were in a non-radar environment, the approach had not been briefed, and the checklists were incomplete. We were focused on the confusion with the FMS, and had allowed ourselves to fall behind. I spoke up saying that I was not comfortable with proceeding directly into the approach because I wasn't 'up to speed' yet. The check airman/PF elected to continue the approach and talked me into it. The radios weren't set and the airplane wasn't even properly configured. We blew through the localizer past full deflection and the PF started using the FMS to plot a course from present position to the runway threshold. The approach was completely unstable and I wanted to call for a go around but didn't want to tell a check airman that. I eventually had to call for a missed approach since we were quite low and way off course. I suggested a climb to the MSA and after some confusion in flying the procedure, we notified ATC.

"We were rapidly cleared for another approach and accomplished checklists late again. We were all searching for situational awareness and spatial orientation. The second approach was also started high and we again overshot the localizer. The autopilot/flight director wouldn't capture the glide slope because we were too far above it and fully deflected on the course. It was at this point that the PF canceling the flight director and flying the approach to minimums using 'raw data.' We were unstable again, I called 'go missed' and we were confused on the missed approach again.

"ATC was notified again and clearance for a third approach was given. This time, we all took a few seconds to relax and calm down. We briefed the approach, set up the radios, configured the aircraft, and completed the checklists. From that point on, the approach and landing proceeded normally with the airplane functioning properly."

This uncomfortable situation was entirely created by poor communication. First, ATC was confused and unclear about the clearance for this crew. Next, the check airman/PF demonstrated extremely poor listening skills, not to mention judgment. Finally, the captain and the FO could have done more to communicate their discomfort to the PF. Throw in the fatigue from a long day and poor weather at the final destination, and you had all the links for a serious mishap. Fortunately, the last link in the accident chain was broken in time.

Pilots rely on ATC controllers. Despite the good-natured ribbing that goes on sometimes and the occasional flare-ups, pilots trust the controllers to keep them safe from hitting something hard—to a point. It is never a good idea to completely trust the very human controller to keep you safe. Keeping you and your passengers safe is ultimately your responsibility. You should always maintain situational awareness with regard to terrain and obstructions on approaches, especially if you are flying below a minimum safe or emergency safe altitude. Although the pilots didn't mention anything in this report, the controller's confusion would have been an immediate warning flag. Any time you are below the minimum safe altitude as depicted on an approach plate, it's a good idea to know the terrain/obstacle that MSA is based on and if you are anywhere near it. If you are near it, query the controller on the min vectoring altitude to make sure he or she takes a second look to help keep you safe. If you seriously doubt your terrain separation at any time, if the hairs on the back of your neck are screaming a warning, request a climb to a more comfortable altitude. If you are really scared, climb first and ask permission after. The same goes if you don't know where you are.

In the case study above, the reporter mentioned that "confusion reigned" on both missed approaches and that the crew was, "all searching for situational awareness and spatial orientation." Above, we talked about when the controller seems lost, but this is a case where the pilot is lost. When that happens, and it does happen, it's critical to regain your situational awareness ASAP. Loss of SA can occur gradually or all at once, but once you recognize that you have lost it, it's relatively easy to recover SA. Tony Kern describes a few steps to recovering SA in *Redefining Airmanship*. First, *get away from "everything hard"* including dirt, rocks, trees, and other aircraft. The key here is the same above. Make sure you are above the MSA and ask for a heading away from other traffic. Next, *stabilize the aircraft*. Level off or otherwise reduce the rate of change to give your gray matter time to catch up. *Buy some time* by requesting a safe heading or even a holding pattern. Once you are stable with some time available, *seek information*. Listen to all inputs. You can even ask ATC for help. This sounds extreme and it is, but if you've lost confidence in yourself or your instruments, you have to restore it before you can continue safely. Finally, *learn from experience*. As soon as possible after losing and regaining your SA, analyze the incident to help with future episodes. Ask what caused you to loose SA in the first place and what cues helped you realize that it was gone. Feeling rushed is a common cue for many aviators including the crew above.

By rushing the first approach, this crew set themselves up for a "where are we" episode. If they would have taken the above steps after the first missed approach, they could have recovered their orientation and probably landed the second time. It was only after the second confusing missed approach that they "took a few seconds to relax and calm down." Those few seconds to stabilize, buy time, and seek information were all they needed to make a successful third approach. After all the stress and anxiety caused by rushing the first two approaches, the third may have even been anti-climactic.

Going back even farther to the beginning of this narrative, and the beginning of the book, the entire approach was poorly planned and never stable. The captain seemed surprised when the check airman took the aircraft. The crew was so focused on the perceived problem with the FMS that little, if any, approach planning occurred. Going through the airmanship model would have definitely helped in this case. And taking the time to go through the model would have given this crew time to get stable and land the first time. What good did it do to rush this approach? Was the check airman in a hurry to get on the ground? A lot of good it did him! By rushing the first two approaches, they certainly didn't get down any earlier.

Finally, there were a host of communication issues on the approaches themselves. The captain didn't want to tell the check airman to go around until absolutely the last minute. It's probably a good thing that it was a captain in the seat instead of the FO. The status difference was far less great, and more easily overcome. How many mishap reports have you read where nothing was said at all? I know I've read far too many. In this case, a classic "halo effect" and a reluctance to challenge a senior flyer prevented timely communication by the captain on two different approaches, to say nothing of the deafening silence from the FO in the jump seat. The reporter doesn't mention that he said anything but he was the least busy person in that cockpit. The jump seater had a clear view of all the instruments and the workload of the two other crew members. Again, the halo effect and reluctance to question an authority figure effectively removed the first officer from the cockpit. The need here for some classic CRM communication tools is obvious.

A well-constructed five-step assertiveness statement, *two-challenge rule,* or *most conservative response rule* would have helped this crew. We've already introduced the five-step assertiveness statement at the beginning of this chapter. Before reading on, think about what you would have said to the check airman if you had been in this situation.* For some reason, the check airman seemed fixated on putting this aircraft down. Perhaps he felt some pressure to show the others how it was really done. Maybe the check airman's ego was a bit too involved in the poor decisions to fly two unstable approaches? Whatever the cause for the fixation, the two-challenge rule is an answer.

The two-challenge rule is a prebriefed arrangement between pilots that the PNF will *automatically* assume the controls if the PF fails to answer two verbal challenges. It was designed by the airlines to guard against pilot subtle incapacitation or fixation. In this case, "Pilot, we are unstable, off course, and off glide slope on this approach. Go around!" If there was no answer after repeating that challenge, then take the controls and go around yourself. If the rule has been prebriefed, much of the anxiety of taking an aircraft from another aviator will be relieved.

Finally, this case study demonstrates many mistakes and their remedies. One of the best, and least frequently used, is the most conservative response rule. From *Redefining Airmanship:*

"Occasionally conflict resolution is ineffective, and airmen need an effective measure to deal with these situations. The most conservative response rule provides a ready solution by simply stating that in the absence of agreement between team members, the default solution will be for the team to "fail-safe" to the more conservative of the options being discussed. The decision to rely

---

*How about, "Pilot, I am extremely uncomfortable and behind the aircraft on this approach. We have not briefed or accomplished checklists. How about requesting a quick hold until we get everything done before heading down? What do you think?"

on the most conservative response rule must occur before the conflict arises, or it merely becomes another part of the argument. Many effective teams brief this rule as a standard operating procedure at the formation stage of the team during preflight briefings.

"Employing the most conservative response rule does not mean that flyers will always be taking the conservative approaches to all situations; that approach is too limiting and inappropriate for the flexibility required in many airborne operations. It simply means that in the absence of agreement in flight, the team will automatically fail-safe to the least hazardous of the options under debate. It is often a difficult step for pilots in command to take, because they perceive it as somehow infringing on their divine right to command. Nothing could be further from the truth. In fact, the most conservative response rule allows the airborne commander more security and freedom to discuss and propose possible courses of action, with the insurance of an assertive crew and a fail-safe tool to ensure reasonableness. Uninhibited by fears of pushing his or her crew into a questionable course of action, the pilot in command is able to use his or her experience, creativity, and leadership to actively address the challenges of the mission."

The most conservative response of the crew above was to take time at the beginning to get the checklists done and get stable before beginning the approach. Let's now look at how GA passengers can also be effective team members and add to the approach safety equation.

Single Pilot and Passenger Communications

Not every pilot has another flyer there to back him up. Many GA pilots fly alone or fly with nonpilot passengers. If you are alone, the only communication that needs improvement is the communication with ATC, but if you are flying with friends or family, you still have communication issues to deal with. As the pilot in command, it is your responsibility to make sure that your passengers don't become a distraction for their own safety. A thorough briefing can go a long way toward making the flight more enjoyable and much safer for the pilot and passengers.

If your passenger is new to flying, try to pick a day that is clear and smooth. If it's turbulent, delay your flight if possible. Flying is difficult enough without having a nervous or panicky passenger next to you. Include your passengers in as much of the flight as possible. Have them join you on the preflight walkaround and explain what you are looking for. Once in the cockpit, point out the various instruments and explain their use. Most GA airplanes have color markings on the instruments showing normal indications. Point those out to your passengers and encourage them to help monitor the instruments and let you know if the needle goes outside the normal indications. Brief them thoroughly on the seat and seatbelt and make sure they can egress the aircraft quickly if needed. Explain the use of checklists and point out the items you are checking. Ask for help in looking for other aircraft on the ground or in the air. Finally, take a tip from your old instructor and talk as you fly. Tell passengers everything you are doing and going to do to help them feel more comfortable. Once they are comfortable, you can cut back on the talking.

And take a tip from the airlines and cut back on the talking during critical phases of flight. The airlines call it the "sterile cockpit" and it's worth implementing with your passengers. The sterile cockpit requires that crew members avoid non-safety-related conversation during taxi, takeoff, and anytime the aircraft is below 10,000 ft except for cruise flight. The same goes for descents, approaches, and landings. Some organizations even require the sterile cockpit when within 1000 ft of your target altitude in climbs and descents. Tell your passengers that for their safety, you want them to avoid idle chatting during certain parts of the flight. Explain that you will be busy listening on the radios for instructions from the controller and for other nearby aircraft. Of course, point out that they are *not* to keep silent about that aircraft off your wing growing steadily larger in the window or that red light. It's just not the time to hear about their last round of golf. Although most of this book revolves around case studies of pilot error, sometimes we learn more from a positive example.

The following case study illustrates many of the communications issues discussed in this chapter as demonstrated by a mature PIC and serves as a summary to our communications chapter. Notice how the PIC continually had to communicate (talk as well as listen) to various people in and out of the cockpit. Also notice how this pilot took time to develop an approach plan and stabilize the approach.

Case Study: Talking to Everyone (ASRS Accession Number 476217; Narrative by SO)

"It's not just the pilots who need to be aware of airspace and NOTAMs [notices to airmen]. It is perhaps more important for the controllers to be intimately familiar with their 'neighborhood.' We filed a flight plan and the captain received the weather and the NOTAMs. The FO copied a normal clearance and we were told to expect a visual approach before being handed off to the approach controller. Once contacted, the approach controller told us that our destination was closed and asked our intentions. I asked for time to discuss options with my passengers as we were assigned a heading. After we evaluated the other nearby airports the controller came up stating that our original destination was, in fact, open and that per ATIS [Automatic Terminal Information Service] and NOTAMs, only Runway 6 was closed.

"We were vectored and cleared for an ILS to Runway 6 and I asked for a go around because we had been expecting a visual approach and weren't ready for the ILS (which was inoperative according to earlier received NOTAMs). The controller then said that the Runway 6 ILS was indeed out of service and asked if we could accept an NDB approach to 6. I said we could but needed a holding pattern for time to prepare. We were given a heading and a climb toward holding when the original approach controller was relieved.

"The new controller immediately started vectoring us for the NDB when we still weren't ready. The controller gave us a heading, altitude, and clearance for the approach but we flew through the approach course. We asked for and received new vectors and finally got established and stable on the approach.

"I asked the FO what altitude we could descend to and he gave me the MDA even though we were still outside the final approach fix. I was looking for an altitude of 3200 ft, not the MDA. We continued with this approach with a circle to land. During this entire time, a passenger was walking around in the cabin trying to stay abreast of the situation despite the 'fasten seatbelt' light being on. I tried to briefly explain what I could, but told her to return to her seat as a full explanation would be available on the ground. We landed uneventfully."

This case had some heavy links in a possible accident chain. ATC was confused and a controller switch during this approach could have complicated things. The FO misunderstood which altitude the captain was requesting and gave one too low for the aircraft's location on the approach. Finally, to add to the confusion in the cockpit, an anxious passenger was out of her seat in the cabin distracting the crew during a busy approach. What broke this accident chain? The answer is good communication.

The PIC was prepared and seemed to know the NOTAM information better than the first controller. He didn't waste time arguing with the controller when the controller originally stated the destination was closed. After all, the controller was unlikely to give an approach clearance to an airport he or she thought was closed. Instead the PIC started discussing other options with the passengers. When the controller acknowledged that the destination was actually open, the captain did not rush the approach but communicated the need for some time to prepare. During the actual approach, the captain's SA alerted him to the wrong altitude called out by the FO and they discussed where they were on the approach and what the appropriate altitude should be. Finally, this captain didn't waste a lot of time and effort talking with his wandering passenger once the approach got busy. He dealt with that distraction efficiently by explaining that a full explanation would be available once safely in the chocks. All in all, this case study is an example of mature and professional communication that dealt with numerous distractions and ensured a safe approach and landing.

Hopefully, this chapter has reminded you of the importance of improving your communication on approaches. Whether you are talking or listening with

ATC or crew members, improving your communication skills will help improve your approaches and your flying in general. When you read case studies, practice the five-step assertive statement you would make if you were the PNF in that situation. How would you challenge the PF if the PF were deviating from standard procedures or the regulations? Evaluate how the individuals in the case studies could listen better and take a hard look at your own listening skills. Could they use improvement? Finally, for the PICs reading this, try briefing and using the most conservative response rule during the next 90 days or 100 h of flying. Get used to using it during day-to-day operations and see if it doesn't encourage more options and freedom during flight. It doesn't diminish your command because *you* have already decided to brief and use the most conservative response. And, if the stuff hits the fan, you'll be ready with a proactive and mature plan.

For the first officers and other crew members, discuss these issues with your captains. Attempt to engage in a dialogue about communication in the cockpit and how to improve it. No, you can't unilaterally implement the most conservative rule, but you can be conversant with the "two-challenge rule" and a five-step statement if the need arises. Someday you'll be in command and one of your responsibilities will be to develop the FOs you fly with. Their skills should be acceptable, but many will not have your experience with good cockpit communication. Help them develop strong "talking" and "listening" skills as the next step in developing their aviation expertise.

In our next chapter, we are going to move away from the human factors associated with approach and landing error and discuss some common errors associated with aircraft performance issues in the airport arrival, approach, and landing environment.

5

Performance Considerations

It has been said that when you violate nature's laws, you are your own judge, jury, and occasionally, your own executioner. Basic physics and aerodynamics tell us that when drag exceeds thrust and lift falls below that which is needed to counteract the earth's gravitational pull, we had better have the landing gear down with a flat surface beneath us. When you are flying an aircraft close to the ground and "dirty" (configured for landing), it is not the time to determine whether or not you have the performance necessary to continue controlled flight to a safe landing. By that time the error has been made, and the outcome is usually unavoidable. Occasionally, it is not a lack of performance that bites us, but too much performance for the runway available.

In each of these cases, the pilot's challenge is to see and feel the performance issue as it arises, and take the necessary steps to match the aircraft performance to the fixed environment (such as runway length) and the changing environment (like wind conditions) to navigate safely within the laws of physics. Judgment comes into play here, but only in the recognition and reactions required to match aircraft to the environment. Physics is an all the time thing, and when we end up on the wrong side of Bernoulli's equation, we are going down.

In this chapter we will look at several aircraft performance issues associated with the approach and landing environment, including aircraft loading, reduced performance considerations, and weather phenomena such as temperature, ice, and tailwind. It is not my intent to spell out preferred techniques or provide "how to" illustrations, but to point out common errors in the hope that you can avoid them in your flying future. Also, astute readers will note that there is no discussion of "cross-controlling" aircraft for alignment or crosswind landings discussed in this chapter, although this technique clearly has performance implications. This subject is addressed in Chap. 10, "Sensory Problems and Landing Illusions." Let's begin with a mishap that involves several elements that, when combined with a high voltage power line, spell tragedy for pilot and passengers.

Case Study: When There Is No "Go" in Go-Around (NTSB Report Number LAX97FA272)

"On August 5, 1997, at 1348 hours Pacific daylight time, a Piper PA28-RT-201, N8227Q, was destroyed when it flew into high voltage power lines approximately two statute miles west-southwest of the Hesperia, California, airport. The private pilot and two passengers were fatally injured. No flight plan was filed for the personal flight, which departed from San Jose, California, at approximately 1130. Visual meteorological conditions prevailed along the route of flight and at the destination airport.

"A witness observed the aircraft execute a go-around from Hesperia airport and noted that the rate of climb appeared slow. The aircraft turned right about 20 degrees and followed a wide depression in the surrounding terrain without any additional gain in altitude. Two witnesses, who were driving along a road that paralleled the aircraft's flight path, stated that the aircraft flew over their vehicle at approximately 100 ft and was observed to jettison three packages, or luggage, out of the right cabin door. Immediately after the third package was jettisoned, the aircraft flew into the power lines and was observed to cartwheel into the ground at which time it caught fire and burned. The witnesses also stated that the engine sounded "normal" and that the aircraft appeared to fly straight and level during the time that it was observed.

"A verbal statement was taken from a witness whose home underlay the final flight path of the aircraft. She stated that she heard the aircraft fly over and heard the sound of the crash, but did not see the aircraft until after it was on fire. She did state, however, that she heard the aircraft engine as it flew over and that it had a smooth, steady sound, with no change in sound level and no other audible noise present.

"The aircraft contacted Edison Company 500,000-volt power lines that traverse the accident site from northwest to southeast approximately 2 statute miles southwest of the runway. The wire impact point was approximately 75–100 feet agl midway between two support towers. There were no orange spheres attached to the power lines at the point of impact.

"At the time of the accident, the weather was reported to be VMC with 30 mi visibility. The temperature and dew point were 104 and 68 degrees Fahrenheit, respectively, and the wind was from 230 degrees at 5 kn. The altimeter setting was 30.05, and the density altitude was 7200 ft msl.

"The aircraft engine was disassembled and inspected with no evidence of internal mechanical failure, foreign object damage, or incorrect assembly. The fuel injection system was clear with no evidence of fuel contamination. The piston faces did not show unusual operating signatures or valve impressions and the rings were slightly worn. The camshaft lobes were not excessively worn. There was no evidence of cracked or burned intake or exhaust valves. It was not possible to establish ignition system continuity due to fire damage to the magnetos and sparkplug leads. All sparkplugs were slightly worn, but indicated normal combustion with a moderately rich mixture when compared to the Champion plug check chart. The exhaust muffler for the No. 2 and No. 4 cylinders had a partially collapsed internal baffle.

"The Textron-Lycoming Engine Operations Manual (part no. 60297-12, page 3.37, figure 3-21, Sea Level-Altitude Performance Chart) indicated that at 7200 ft density altitude, the engine was capable of developing approximately 72 percent of its rated power.

"The sheriff's deputies who responded to the accident retrieved the objects that were thrown from the airplane prior to the accident. They stated that the objects were three pieces of luggage containing the personal effects of the deceased. They estimated the total weight of all three to be less than 100 pounds."

The National Transportation Safety Board determines the probable cause(s) of this accident as follows: "failure of the pilot to see-and-avoid high voltage power lines, while coping with a lack of engine power during a go-around in a high-density altitude condition. Factors relating to the accident were: high-density altitude that resulted in a normal reduction of engine power; partial obstruction in the exhaust system that resulted in an additional loss of power; and the encounter with power lines."

Analysis

There are many unanswered questions with this particular mishap, such as why was the go-around attempted? Did the pilot know of the problems with the internal baffles? Or had he checked the density altitude of the field against the aircraft's performance charts? But each of these questions provides clues to preflight preparation steps that might have avoided this accident. For example, there are times when you can predetermine that a go-around is not possible at the density altitude you are flying into. When this occurs, the decision to proceed anyway turns into a one-way decision gate—a situation that in and of itself is dangerous enough, but becomes more critical if the one-way gate is not recognized, as I suspect in this case, and in many other performance-related mishaps that I have reviewed over the years.

This performance awareness issue also ties into the in-flight options that might have been available to the pilot when the go-around was attempted. First, if we assume that the pilot was unaware of the performance inadequacies of his aircraft in the high-density altitude, he took logical steps by turning toward an area of lower terrain when he realized that he had a problem and again made a valiant effort by reducing gross weight by throwing out the luggage. These steps may well have been adequate, if it wasn't for the power lines. Even then, the pilot might have survived if he had seen the power lines in time to go under them, but they were unmarked because they weren't along a regular flight path.

From this case study, you can begin to see how far back a mishap chain of events extends with regards to performance criteria and pilot decisions. Perhaps this mishap began with a lack of preflight planning, perhaps even further back with the failure of the baffles in the exhaust systems that went undetected. Certainly the event began in earnest with the decision to execute a go-around at a density altitude of 7200 ft in an area of high terrain at 72 percent of normal power in a heavily loaded aircraft. But it isn't only hot weather and high-density altitude that can cause performance problems. Trouble lurks at the other end of the weather spectrum as well.

Back to the Basics: Thrust, Drag, Lift, Weight

Airfoils are magnificent evidence of the creative capabilities of man. Thousands of years of watching the examples of birds, followed by thousands (perhaps tens of thousands) of failed experiments were required for man to reach the stage where geniuses like da Vinci, Gerard, and Meerwein would begin to design and construct the delicate and exact curves of the ornithopter wing.

Later, the Wright brothers and other inventors would combine the technology of the internal combustion engine and the lifting surface to produce what man had dreamed about for thousands of years—controlled, sustained flight. How odd that modern flyers have forgotten how fragile and critical the curvature of the manufactured wing really is. One example of this curious phenomenon follows, where a pilot approaching an Ohio airport finds himself in a world of trouble, and it shouldn't have come as a surprise.

Case Study: Ice and Performance (NTSB Report Number NYC96FA085)

"On April 16, 1996, about 1028 eastern daylight time, a Piper PA-28-181, N8276Y was destroyed during a forced landing when it impacted in an open field near Granville, Ohio. The instrument rated private pilot was fatally injured and the passenger received serious injuries. Instrument meteorological conditions prevailed for the flight that departed the Merrill C. Meigs Airport, Chicago, Illinois, at 0805, destined for Athens, Ohio. There was an instrument flight rules flight plan for the personal flight conducted under 14 CFR Part 91.

"The pilot proceeded en route at 9000 ft and was in contact with Chicago and Indianapolis centers. The pilot was in contact with Columbus approach when he reported a partial loss of engine power. Several vectors to nearby airports were provided to the pilot; he was being vectored toward the Newark-Heath Airport, Newark, Ohio, when the accident occurred.

"According to the passenger, the flight was delayed one day due to weather, and that she was scheduled to attend a class, at the destination, the day of the accident. Prior to departure, the pilot advised her that the weather forecast called for isolated icing; however there were no pilot reports of those conditions. Also, they could get occasional icing, and he reassured her that they could get out of those conditions immediately. When they departed Chicago, it was very foggy, and cold. Additionally, that they were in foggy conditions for the entire flight.

"The pilot received a weather briefing by telephone from the Kankakee Flight Service Station (FSS) about 0645. Review of the rerecorded briefing revealed the pilot was advised, in part, of icing conditions en route. According to the rerecording, the FSS Specialist stated: 'Okay everything looks good through Illinois and Indiana, it's just into Ohio where you run into advisories for icing and turbulence...specifically occasional moderate rime or mixed icing basically above two thousand up to thirteen thousand feet....' The briefer provided the pilot with a pilot report, and stated, 'on descent into O'hare, an ATR 42...commuter airliner reported light rime icing niner thousand down to eight thousand, didn't say anything about lower than that. Pretty tough conditions for a PA twenty-eight.' The briefer later stated about the destination area: 'So it's good enough to get in—if you couldn't get in...you got some good alternates up toward Columbus in that area...so I don't think you have a problem there, it's just my main concern on this, on this one's gonna be ice.' The pilot then stated 'Yeah, but it sounds like once I get out of Chicago..., the only pilot report you got on that is higher up.' At which time the briefer stated, 'I've got six thousand foot temperatures for that entire route ranging from minus six to minus eight degrees [Celsius]...I don't have any top reports either, that's the problem.' The briefer later provided the pilot with the winds aloft and stated: 'Your winds aloft out of Chicago, you're gonna have a good tail wind there's no doubt about that...out of Chicago at three thousand from three one zero at thirty-two, six thousand uh same thing, temperature minus eight [degrees Celsius], niner thousand from three one zero at three six, temp minus twelve [degrees Celsius]. Indianapolis, they read the same as Chicago. Columbus, at three thousand from two seventy at two four, six thousand from two ninety at two niner, temp minus seven [degrees Celsius], niner thousand from two eight zero at three one, temp minus twelve [degrees Celsius].'

The briefing ended about 0700, after the pilot filed an IFR flight plan, and the briefer made another check for pilot reports, with none found.

"The pilot had been communicating with the north radar controller (NRC) at Columbus Approach Control when he reported engine trouble and requested vectors to the nearest airport. A review of the Columbus Approach Control communication tapes and radar data revealed the following: The pilot was en route at 9000 feet, about 13.6 miles northeast of Columbus, Ohio, and 12.5 mi southwest of Mount Vernon, Ohio, when he radioed, at 1016:33, 'Columbus approach, this is eight two seven six yankee uh...engine trouble.' NRC asked the pilot at 1016:47, '...what can I do for you right now?' The pilot responded, at 1016:54, 'Vectors to the nearest airport if I could.' When questioned by the air traffic controller if he lost total engine power, the pilot responded '...we're developing a little bit of power....' At 1016:56, AC provided radar vectors initially to the Knox County Airport, Mount Vernon, Ohio, and at 1017:30 told the pilot '...descend and maintain three thousand that is at pilot's discretion on the altitude.' At 1017:44, NRC provided the pilot with the weather at Port Columbus International, Columbus, Ohio, and stated: '...your presently one three miles north east of Port Columbus if you want to try there.' The pilot radioed that he would take Port Columbus. At 1019:01, the pilot asked 'How far from the airport now?' NRC responded 'Cherokee seven six yankee now one five miles northeast of the airport.' AC then radioed at 1019:18, 'Cherokee seven six yankee published minimum descent altitude is one thousand three hundred and forty feet presently descend to three thousand feet.' The pilot was then given a frequency change to the final radar controller.

"The final radar controller (FRC) advised the pilot, at 1020:27, that he was one three miles northeast of Port Columbus, and asked if he checked the carburetor heat, which the pilot responded, 'yeah I did, it's not working.' The pilot then asked 'Anything closer than Port Columbus?' The FRC radioed at 1021:03, that Knox County Airport was northeast, which would give him a tailwind and provided a 040-degree heading. At 1021:21, the FRC asked the pilot to fly a 130-degree heading. When the pilot asked, at 1021:49, '...how far from the airport now,' FRC responded, '...your eleven miles, one one miles northwest of the Newark Airport.' FRC radioed at 1024:37, 'Cherokee seven six yankee, I have ah reports at Columbus of light rime icing all the way down to the final for Columbus.' At 1025:54, the pilot asked '...how far from the airport and the wind direction please,' to which FRC responded that he was 6½ miles northwest.' No further transmissions were received from the pilot.

"The surface observation for Columbus at 1008, was as follows: measured ceiling 1300 broken; 1900 overcast; visibility 2 mi with light rain and fog, temperature 38 degrees F; dewpoint 36 degrees F; altimeter 29.75″ Hg.

"The airplane came to rest in an open field. According to the Granville Fire Chief, he arrived at the accident site about 1050, *and stated that he observed pieces of ice, about ½″ thick, in the vicinity of the wreckage, conformed to the shape of the leading edges of the wings*" [emphasis added].

Airplane Limitations

The pilot's operating handbook stated, in the limitations section, for types of operation: "The airplane is approved for the following operation when equipped in accordance with FAR 91 or FAR 135. (a) Day V.F.R. (b) Night V.F.R. (c) Day I.F.R. (d) Night I.F.R. (e) Non Icing."

Icing Information

The Advisory Circular, "Pilot Precautions and Procedures to Be Taken in Preventing Aircraft Reciprocating Engine Induction System and Dual System Icing Problems," under impact ice, stated:

"Impact ice is formed by moisture-laden air at temperatures below freezing, striking and freezing on elements of the induction system which are at temperatures of 32 degrees F [0 degrees C]. Under these conditions, ice may build up on such components as the air scoops, heat or alternate air valves, intake screens, and protrusions in the carburetor. Pilots should be particularly alert for such icing when flying in snow, sleet, rain, or clouds, especially when they see ice forming on the windshield or leading edge of the wings. The ambient temperature at which impact ice can be expected to build most rapidly is about 25 degrees F [−4 degrees C], when the supercooled moisture in the air is still in a semiliquid state. This type of icing affects an engine with fuel injection, as well as carbureted engines. *It is usually preferable to use carburetor heat or alternate air as an ice prevention means, rather than as a deicer, because fast forming ice which is not immediately recognized by the pilot may significantly lower the amount of heat available from the carburetor heating system. Additionally, to prevent power loss from impact ice, it may be necessary to turn to carburetor heat or alternate air before the selector valve is frozen fast by the accumulation of ice around it. When icing conditions are present, it is wise to guard against a serious buildup before deicing capability is lost*" [emphasis added].

The Aeronautical Information Manual for pireps (pilot reports) relating to airframe icing stated:

"The effects of ice on aircraft are cumulative—thrust is reduced, drag increases, lift lessens, and weight increases. The results are an increase in stall speed and a deterioration of aircraft performance. In extreme cases, 2 to 3 inches of ice can form on the leading edge of the airfoil in less than 5 min. *It takes but ½ inch of ice* (exactly the amount noticed by the Sheriff at the crash site) *to reduce the lifting power of some aircraft by 50 percent and increases the frictional drag by an equal percentage* [emphasis added]. A pilot can expect icing when flying in visible precipitation, such as rain or cloud droplets, and the temperature is 0 degrees Celsius or colder."

Gliding Distance?

"The airplane was approximately 12.5 mi southwest of Knox County Airport, about 9000 feet, with a westerly wind at about 30 kn, when the pilot requested vectors to the nearest airport due to a partial loss of engine power. The pilot later radioed that he was holding his best glide speed. According to the Pilot Operating Handbook, with the conditions above, the airplane was within gliding distance to the Knox County Airport (under normal non-icing conditions)." In effect, the ice not only took away the power and lift necessary to sustain flight, but also the last hope of a pilot—the ability to glide to a safe forced landing location.

The National Transportation Safety Board determines the probable cause(s) of this accident as follows:

"The pilot's intentional flight into known icing conditions and his overconfidence in his ability, which led to a partial loss of engine power as a result of induction system ice. Also causal was ATC's excessive vectoring of the airplane in icing conditions, further deteriorating the airplane's performance, and thereby placing the airplane beyond gliding distance to a nearby airport. In addition, the pilot failed to maintain airspeed during the forced landing, which resulted in a stall. The icing conditions were a factor."

Analysis

This example makes clear the point that once ice begins to accumulate, it is often too late to reactively create a safe outcome—another one of those one-way decision gates. This pilot had flown into known icing conditions, a situation prohibited by the flight manual, and then likely failed to turn on the carb heat prior to the initial icing buildup. According to the AIM, this could have reduced his lift capability by 50 percent, while continuously adding to the weight of the aircraft, although we will never know whether the pilot was unaware of the icing restrictions or procedures, or if he just forgot under the pressure of the moment. What is clear, is that he failed to adequately use this knowledge in a timely manner in flight. The lessons from this tragedy are clear:

1. Know the limitations of your aircraft.

2. If you find yourself approaching *potential* icing conditions, get your anti-icing equipment and carb heat on to prevent buildup before the capability is lost.

3. Communicate the severity of your situation clearly to the air traffic controller.

4. Remember, time is your enemy here. If your aircraft is not equipped to handle the icing conditions you have encountered, you must exit these conditions as quickly as possible.

Every year, this sequence of events plays out over and over again in small aircraft not equipped for icing conditions. Yet many pilots will find themselves flirting with an icing encounter, counting on their in-flight risk management skills and air traffic control assistance to get them out of ice if they encounter it. Far too often, this is a fatal flaw.

He Should Have Seen It Coming (NTSB Report Number CHI98FA107)

On March 20, 1998, at 2005 eastern standard time (all times EST), a Beech A-36, N1812A, was destroyed when it impacted terrain 5 mi south of the Sullivan County Airport, Sullivan, Indiana. The pilot reported picking up ice and was being vectored for an instrument approach. The instrument-rated private pilot, pilot-rated passenger, and two passengers received fatal injuries. The 14 CFR Part 91 business flight departed Louisville, Kentucky, en route to Aurora, Illinois. Instrument meteorological conditions prevailed and the aircraft was on an instrument flight plan. The following sequence illustrates what can occur when you flirt with a known killer.

"At 1818, the pilot called the Louisville Automated Flight Service Station (AFSS) by telephone and obtained a preflight pilot briefing for an IFR flight from Louisville, Kentucky, to Aurora, Illinois. The pilot indicated to the weather briefer that he was concerned about the icing conditions. The weather briefer reported the following flight precaution: 'Okay, there is an Airmet for occasional to moderate rime or mixed icing along the entire route of flight. Ah, 3000 to 16,000 here locally. Up north, from the surface to 16,000 up there in the Aurora area. And, Airmet for an occasional IFR conditions along the entire route of flight. Turbulence, ah, occasional moderate below 8000. That's it on the flight precautions. Low pressure currently located in…extreme eastern Kentucky. It's moving slowly to the east. As it does, it should pull some of the moisture with it, but, what you'll be encountering affecting your route of flight is the wrap-around effect on that low pressure.'

"The AFSS weather briefer gave the pilot reports concerning icing conditions encountered by airplanes near Louisville, Kentucky. A Cessna 195 had reported light to moderate rime icing that was shed when he descended from 4000 feet to 3000 ft mean sea level (msl). A Cessna 340 had reported moderate to severe mixed icing at 6000 feet msl. An Aero Commander pilot had reported light rime icing at 7500 ft msl near Lafayette, Indiana. No pilot reports of icing had been received from pilots in the Chicago, Illinois, area.

"At 1852, the pilot called the Louisville AFSS by telephone and filed an IFR flight plan from Louisville, Kentucky, to Aurora, Illinois. The briefer asked the pilot if he needed a weather brief and the pilot declined.

"At 1917, the pilot departed Standiford/Louisville International Airport, Louisville, Kentucky, en route to Aurora, Illinois. At 1918, the pilot was instructed to climb to the filed en route altitude of 4000 ft msl. At 1921, the pilot was instructed to turn right on course. At 1941, the pilot was instructed to contact the Evansville Approach Control.

"At 1854, Evansville Approach Control had received a pilot report from a Cessna 402. It had encountered light mixed icing about 30 mi north of Evansville, Indiana, at 8000 feet msl and that it lost the ice at 4500 ft msl.

"At 1941, N1812A (12A) contacted Evansville Approach Control and reported level at 4000 ft msl.

"At 1942, the pilot requested icing reports in the area. Evansville Approach reported negative icing reports in the area.

"The controller reported that, 'The aircraft [12A] was in the far northeast corner of my airspace and I had no reports in that area.'

"At 1945, Evansville Approach Control told 12A to contact Terre Haute Approach Control.

"At 1945:37, 12A contacted Terre Haute Approach Control and reported level at 4000 feet msl.

"At 1945:58, 12A requested icing reports in the area.

"At 1946:01, the controller responded, 'November One Two Alpha, I haven't had any in several hours. Um, earlier they were saying at or below five with some light to moderate mixed. The last one we had about two hours ago was, um, descending into Terre Haute. Said the ice they picked up was between ten and eleven.'

At 1946:17, 12A responded, 'Two Alpha, thanks.'

"At 1946:23, the controller stated, '...If you have any pilot reports for me, I appreciate it. And, um, just let me know if you need to deviate for anything.'

"At 1949:13, 12A requested a descent to 3000 ft msl and was cleared to descend at 1949:25.

"At 1952:48, 12A reported, 'Yah, we're starting to accumulate some ice. Don't think were gonna shed it. Any chance of, uh, getting Terre Haute tonight?'

"At 1952:54, the controller reported, '...sure, you can fly heading three six zero toward Terre Haute. Ah, I'll have weather for you shortly.'

"At 1955:30, the controller reported, '...the Terre Haute weather: wind three six zero at one one, gust to one seven, visibility one zero, ceiling is nine hundred, broken one thousand four hundred, overcast. Altimeter two niner seven zero. And fly heading three four zero. Vectors for the ILS runway four.'

"At 1955:56, the controller reported, '...if you like you can descend at pilot's discretion two thousand five hundred.'

"At 1956:01, the 12A responded, 'Down to twenty-five hundred, One Two Alpha.'

"At 2001:48, the 12A reported, 'Is there any where else (it seems) we're losing a little more airspeed here and, ah, I don't know anything closer than Terre Haute.'

"At 2001:54, the controller reported, 'Well, at your twelve o'clock, eleven or twelve o'clock, and five miles is Sullivan Airport.'

"At 2002:00, 12A responded, 'You think we can get into that?'

"At 2002:04, the controller responded, 'Um, yah, if you'd like. Two thousand five hundred is as low as I can get you. Ya IMC?'

"At 2002:08, 12A responded, 'Yah, we still IMC here at twenty (unintelligible).'

"Between 2002:10 and 2004:07, 12A was given vectors for the VOR/DME-A approach into Sullivan County Airport, Sullivan, Indiana.

"At 2004:40, the controller stated, 'And, Bonanza One Two Alpha, about, ah, two and a half miles from Rodeo. Turn right heading zero two zero, maintain two thousand five hundred till established on the VOR final. Cleared DME Alpha approach.'

"At 2004:51, 12A responded, 'Zero two zero (unintelligible).'

"At 2005:42, the controller stated, 'Bonanza One Two Alpha, Hulman.'

"At 2005:45, 12A responded, 'Heck, we're all over the place right now for One Two Alpha.'

"At 2005:47, the controller stated, 'One Two Alpha, roger, and if you can fly heading of three two zero when able, and ah, climb and maintain two thousand five hundred, we'll try to get you back on there. Is that, ah, wind that low pretty strong?'"

There were no further radio transmissions from N1812A. The airplane was found about five miles south of the Sullivan County Airport. There were no witnesses to the accident. The Beechcraft Bonanza A36 Flight Manual's Limitation Section contained the following warning: FLIGHT IN KNOWN ICING CONDITIONS IS PROHIBITED.

Many pilots reading this account will find themselves uttering (as I did), "There but by the grace of God go I." Some will say that the pilot did all that could be reasonably expected, and that he was just dealt a bad hand on descent. This is also partially true. The pilot did take many precautions, and that is the true value of this example. It only takes once and all of this pilot's precautions did not keep him from flying into icing conditions. Ice is a killer of small aircraft on approach and landing. Don't forget it. Stay vigilant and be prepared when the best precautions fail.

There is a little bit more to the aircraft performance story than just weight and lift, and the discussion revolves around a small point on the aircraft called the "center of gravity."

Weight and Balance Issues in Flight Performance

In order to fully understand performance issues as they relate to lift and weight, it is important to grasp the role of the fulcrum around which these forces play, usually referred to as the "center of gravity" or simply "CG." Aircraft behave differently when the weight is distributed differently. An aircraft with weight distributed more toward the back, commonly referred to as having "an aft CG," makes the airplane less stable, and recovery from maneuvers somewhat more tricky. However, with an aft CG, the airplane stalls at a slightly lower airspeed. To counteract the tail heaviness of the aft CG, the elevator must be trimmed for an upload. The horizontal stabilizer, as a result, produces extra lift and the wings, correspondingly, hold a slightly lower angle of attack.

An airplane with a forward center of gravity feels somewhat nose heavy. It is more stable but requires more backpressure or trim in order to raise the nose—a fact to remember when flying near the performance margins, such as during takeoffs and in the landing flare. The forward CG means a somewhat higher stalling speed, another fact to remember during takeoffs and landings. These are subtle but important differences and the thing to remember is that these characteristics are more pronounced as a pilot approaches the fore and aft CG limits and can become dangerous—even deadly—if they are exceeded.

So far, most of our discussion has been about issues of degraded performance, but high risk is also present at the other end of the performance spectrum, when we have too much energy and the challenge is to get rid of it before it takes us somewhere we never intended.

Too Much of a Good Thing?

Occasionally, there are times when too much—rather than too little—performance is the problem. Every year, mishaps occur where aircraft attempt to land too heavy or too fast, and end up sliding off the end of a prepared surface. The key to safe operations is to *know*—not guess—what your stopping distance is, and allow room for a few surprises if you can. Bruce Landsberg, a pilot and an aviation safety expert who writes prolifically for AOPA's excellent Safety Pilot series, cites several common errors that are made by pilots who end up "off the end and in the dirt." Let's briefly look at each one.

- *Too heavy.* As difficult as it is to believe, there are many pilots who willfully overload their aircraft. They cite many justifications, such as "as soon as we burn off fuel, it will be all right" or "there is a safety margin built into the performance charts."

- *Improper configuration.* Some pilots seem reluctant to drop a notch of flaps for a short field situation. This can, and often does, result in a high-energy condition that becomes critical with a tailwind or on a slippery runway.

- *Too fast.* There is an old pilot adage that says to "carry a few extra knots for the wife and kids." Certainly, no one would advocate getting below approach speed, but in many cases, too much is as bad as too little.

- *Bouncing.* Wheel brakes are only effective in slowing you down when the wheels are on the ground. Sometimes a pilot can attempt a "firm landing" to assist in dissipating energy, and end up floating down the runway after a bounce.

- *Landing too far down the runway.* All of the above conditions can result in a long landing, obviously complicating the braking and stopping equation.

- *Improper go-around decision.* Frequently, the last mistake in a landing accident mishap chain is the pilot's unwise decision to attempt a go-around with insufficient runway available.

Take a look at the following case studies and see how many of these errors you can identify.

Can You Stop? (NTSB Report Number ATL97LA074)

"On May 22, 1997, at 1745 eastern daylight time, a Piper PA-28-151, N8392C, collided with trees several hundred feet north of the departure end of Runway 03 at the Johnson Airport in Sophia, North Carolina. The business flight operated under the provisions of Title 14 CFR Part 91 with no flight plan filed. Visual weather conditions prevailed at the time of the accident. The airplane sustained substantial damage, and the pilot was not injured. The flight departed Hickory, North Carolina, at 1700.

"According to the pilot, while flying at 3000 ft en route to Raleigh, North Carolina, the engine rpm dropped from 2200 to approximately 700. Since level flight could not be maintained, the pilot selected a nearby private airstrip and attempted an emergency landing. The pilot said that the airplane bounced as it touched down on the sod runway, and he noticed an increase in engine rpm. The pilot also stated that he realized that he could not get the airplane down before the end of the runway, and decided to go around. The airplane climbed to the height of the trees, before it settled into the trees.

"Examination of the airplane disclosed that electrical tape was lodged in the carburetor induction system. The engine examination also disclosed that someone had wrapped black electrical tape around the flexible duct material between the induction air filter and the carburetor. A review of the aircraft maintenance logs failed to disclose when or who performed the repair on the air duct. A review of approved aircraft maintenance procedures failed to disclose a repair procedure that would allow the use of electrical tape to repair the air duct. A review of normal aircraft performance data disclosed that approximately 600 ft of ground roll is needed for a landing; approximately 1200 ft of runway is needed to land over a 50-foot obstacle. The pilot reported that the sod runway was 2070 ft long."

The National Transportation Safety Board determines the probable cause(s) of this accident as follows:

"Improper maintenance performed by an unknown person, which resulted in the reduction of airflow through the engine induction system and subsequent loss of engine power. A factor related to the accident was: failure of the pilot to attain a proper touchdown point on the runway for a successful forced landing."

In defense of this pilot's actions, he was attempting to land at a field he was unfamiliar with. Because it was a small plane, with a single pilot on board, this mishap might easily be tossed into the "who cares" category. But take a moment to compare this case study with the one that follows, and see if you can't identify some commonalities and begin to appreciate the hazards of "too much."

Late Descent, Tailwind, and the "Same Old Story" Writ Large

"All I can do is verify that the pilots have been terminated," the Southwest Airlines spokeswoman said. "We can't give any other details because the National Transportation Safety Board investigation is still pending." With those words, two otherwise excellent pilots saw the end—or at least what looks like the end—of otherwise highly successful careers. The spokeswoman was referring to Southwest Flight 1455, a Boeing 737-300, arriving from Las Vegas, that skidded off Runway 8 at Burbank onto a city street March 5, striking a car, just missing a gas station and slightly injuring 15 of the 142 people on board.

This was a hugely embarrassing event for any airline, even making the *Tonight Show* monologue, with Jay Leno remarking, "The aircraft came to rest at a Chevron gas station in Burbank...but the passengers' luggage still ended up at a Shell station in Ontario!" Funny stuff for Leno's audience, not funny for the pilots or executives at Southwest. How could this happen with an experienced flight crew on a routine mission with all equipment operating on a clear night?

The NTSB reviewed flight recordings, radar data, and interviews with air traffic controllers from the accident. Investigators said the plane made its final descent into Burbank-Glendale-Pasadena Airport at more than a 6-degree angle, double the Southwest "stabilized approach" criteria that are required at 500 ft AGL if the landing is to be continued. Because of the steep angle, the crew was unable to control the airspeed going into the 6000-ft runway and landed nearly 40 kn hot. "At that speed, they never stood a chance of staying on the runway," said aviation consultant C. O. Miller, a former senior investigator for the NTSB.

A review of the mishap identified that the crew may have gotten caught up in a rushed approach. Not an excuse, but perhaps one explanation as to how an 18,000-hour pilot flying for what was statistically the world's safest airline could make such an error in judgment. Arriving in Burbank on a flight from Las Vegas, Nevada, the crew contacted approach control and was told to expect the visual approach to Runway 8 at Burbank. Approximately 10 mi from the field, while descending to 3000 ft, the crew was instructed to maintain 230 kn until further advised. One minute later, the crew was cleared for the visual approach to Runway 8, with an instruction to maintain 3000 ft until passing the Van Nuys VOR (approx 6 nm from the runway). The crew passed the VOR and failed to start its descent from 3000 ft. Approximately 4 nm from the runway threshold, at an airspeed of 230 kn and an altitude of 3000 ft, the crew began its descent to land. Complicating the approach was that fact that there was a 7-kn tailwind on the field for landing. Because of the steep nature of the descent (nearly 7°), the crew received two "sink rate" warnings at approximately 400 ft AGL, and a "pull up" warning at 190 ft AGL from its automated systems. The aircraft touched down 2800 ft down the 6032-ft runway with a ground speed of 181 kn. At that point, the crew's fate was sealed.

Despite using all available stopping systems (max reverse thrust, spoilers, and brakes) the crew was unable to stop the aircraft before the end of the runway, broke through a blast fence at approximately 40 kn, skidded across Hollywood Way, and came to rest 38 ft from a Chevron gasoline station. The aircraft was evacuated via the escape slides.

This is not the only recent event where winds and approach speed resulted in a major airliner going off the end of the runway. American Airlines Flight 1420 landed during a thunderstorm in Little Rock, Arkansas, and slid 5000 ft before hurtling over an embankment and breaking apart on heavy metal light poles. Eleven people, including the highly experienced captain, were killed. Too much energy, when coupled with too little time and distance to dissipate it, is a recipe for disaster.

We have looked at aircraft performance considerations with regards to weather phenomena and excess performance because of improper pilot technique and tailwind factors. Let's now look briefly at another performance issue one does not usually associate with the approach and landing environment—pilot-induced stall/spin due to overaggressive maneuvering.

Aggressive Maneuvering

Pilots are typically high-performing individuals. We fly for a variety of reasons, but most pilots like to think of themselves as "high-speed low-drag" performers in and out of the cockpit. I like my car to run as well as it can, and have been known to purchase an octane booster gasoline additive on occasion. My point is simply that high performers like to squeeze every ounce from life, and this can set us up for the big sleep if we push an aircraft past its mechanical or aerodynamic limits, as seen in the following example.

Case Study: Trying to Get All He Could Out of the Machine (NTSB Report Number FTW97FA274)

"On July 19, 1997, at 1925 central daylight time, a Hillam Rans S-9 experimental airplane, N1678N, was destroyed following a loss of control while attempting to return to the airport after experiencing a loss of engine power during initial takeoff climb near McKinney, Texas. The student pilot, sole occupant of the homebuilt airplane, was fatally injured. The airplane was owned and operated by the pilot under Title 14 CFR Part 91. Visual meteorological conditions prevailed for the personal flight. The airplane departed from the Aero Country Airpark approximately 5 min prior to the accident.

"Witnesses at the airpark reported that the airplane executed a high-speed low pass over Runway 17 at an estimated altitude of 4 ft above the runway. About ¾ down the 2950-ft runway, the pilot initiated a 'sharp pull up' establishing a pronounced nose high attitude. Witnesses further reported hearing that the engine suddenly lost power, and they observed that the 3-bladed propeller had stopped at an estimated altitude between 350 to 400 feet AGL.

"Witnesses further stated that the pilot established a glide as he turned to a northerly heading toward the runway. Witnesses further stated that an ultralight vehicle that had been holding short of the active while the low pass was completed by the Rans proceeded to taxi into position for takeoff.

"Another witness observed that the ultralight vehicle was on his takeoff roll as the Rans was gliding northbound 'as if he was trying to sequence himself behind the ultralight on the runway.' Witnesses at the airport continued to watch the airplane 'expecting a downwind landing' to either the runway, the taxiway, or the grassy area east of the runway, but instead, the airplane continued gliding northbound on a close left downwind down to an estimated altitude of 50 to 100 ft AGL.

"Witnesses reported that 'as soon as the ultralight was past him, the Rans appeared to try to turn to get into the wind by making a 180-degree left turn in behind the ultralight.' The witness added that 'he was so close to the runway that it had to be a very tight turn.' The witness further stated that while executing the left turn 'the wing dropped and the nose turned and dropped simultaneously.' At that point the witnesses observed the airplane 'rotate ¼ to ⅜ of a complete spin' prior to the airplane impacting into the trees in a near vertical attitude.

"The pilot was reported to have started to fly around 1992. He purchased an ultralight vehicle and 'taught himself to fly.' According to a friend, the pilot had accumulated approximately 300 flight hours since 1992. His second airplane was another experimental airplane, which he bought used from the original builder. The accident airplane was the third airplane he owned, and like his previous airplane was also purchased from the original builder in Idaho.

"Personnel at the airport reported to the NTSB investigator-in-charge (IIC) that the pilot had previously owned one ultralight vehicle and another home built airplane prior to purchasing the accident airplane about a year earlier. The pilot had performed several modifications to 'improve aircraft performance' and was currently in the process of 'experimenting' with the carburetor and fuel system to increase the power output of the installed Rotax 670 engine.

"The pilot held a student pilot certificate number with a third class medical certificate issued on August 24, 1994. The pilot's logbook was not located. The student pilot, who was employed as an automotive technician specializing in fuel injected fuel systems, had recently replaced the original exhaust system with a high-performance exhaust system. Prior to the accident flight, the pilot replaced the stock carburetor with a mechanical fuel injection system."

Analysis

Put in its simplest form, this pilot pushed his aircraft and his piloting skills beyond their limits. When the engine failed, there was still every opportunity for a safe landing, but by attempting to sequence himself after the ultralight, coupled with a miscalculation on runway displacement that required a steep turn to final, the pilot got on the wrong side of Bernoulli's lift equation.

This case study brings up another critical point for discussion—preparing for the unexpected. Part of the decision matrix faced by the previous pilot was the poor timing of the engine failure (high-speed pull-up close to the ground). Clearly this could not have happened at a more inopportune moment, and the subsequent actions were significantly influenced by the surprise of the event. Sometimes, however, there is adequate time to prepare for the unforeseen, such as during a pretakeoff briefing, as in our next case study.

Ready for Takeoff!...Ready to Land? (NTSB Report Number MIA98FA250)

"On September 20, 1998, about 1431 eastern daylight time, a Piper Aerostar 600, N17MT, registered to a private individual, crashed shortly after takeoff from the Grand Strand Airport, North Myrtle Beach, South Carolina. Visual meteorological conditions prevailed at the time and an Instrument Flight Rules (IFR) flight plan was filed for the 14 CFR Part 91 personal flight. The airplane was destroyed by impact forces and a postcrash fire. The commercial-rated pilot and three passengers were fatally injured. An individual on the ground who was injured by fire subsequently died 6 days after the accident. The flight originated about 3 min earlier.

"According to a line service person employed by Ramp 66, North Myrtle Beach, the airplane had been hangared since arrival 3 days earlier and before the flight departed, he noted that the pilot was seated in the left front seat and a younger looking person of the three passengers was seated in the right front seat. Both engines started "ok," and the pilot waited on the ground for about 20–25 minutes with the engines idling. During that time he observed the pilot wiping sweat from his face using a cloth; he was doing something on the panel, and he put on his headset and kneeboard. He reported that he drove a golf cart to a corner of the field for the purpose of watching the takeoff, and while there, the airplane arrived at the run-up area. *He did not hear an engine run-up, and noted that the engines remained at idle while the airplane was stopped there for about 5–6 minutes* [emphasis added]. He then positioned his cart to a point about midfield of the runway to observe the takeoff; the airplane departed with no flaps extended. During the climbout, he observed the landing gear retracting and a slight amount of black smoke trailing the airplane.

"Review of a transcription of communications with the Grand Strand Airport Air Traffic Control Tower revealed that at 1420:24, the pilot contacted ground control, requested and received his IFR clearance, then at 1421:07, the controller cleared the flight to taxi to Runway 23. The pilot was provided the wind and barometric setting information. At 1427:34, the pilot radioed the local controller and advised that the flight was ready for departure on Runway 23. The controller cleared the flight to takeoff at 1427:39, which was acknowledged by the pilot 6 seconds later. At 1428:32, the local controller advised the controller from the arrival position of Myrtle Beach Approach Control that 'aerostar seven mike tango's rolling....' At 1429:19, the local controller radioed the pilot and advised him 'seven mike tango appears to have smoke trailing from your right side.' The pilot acknowledged the transmission and at 1429:37, the pilot radioed the local controller and stated 'seven mike tango we're going to come around and land.' The local controller acknowledged the pilot's transmission and asked the pilot if he was declaring an emergency or did he need any assistance. The pilot replied negative. At 1429:58, the local controller cleared the flight to land on Runway 23, which was acknowledged by the pilot. Further transmissions from the pilot recorded on the voice tape were unintelligible. At 1430:53, the local controller advised the approach controller that the airplane had crashed and smoke was observed.

"The controller who observed the smoke reported in a personnel statement that '...as the aircraft was passing in front of the tower it appeared to be trailing smoke. I picked up the binoculars and confirmed my observation and identified the smoke as coming from the right engine and smoke was beginning to increase in intensity. I advised the pilot (he was crossing the departure end of Runway 23) that the right engine appeared to be trailing smoke....' He later stated that the smoke appeared to be deep gray in color with two dark edges that appeared to be 1 to 1.5 feet apart with distinct separation. He did not see any fire or flames.

"Numerous witnesses near the accident site reported observing a low-flying airplane with smoke trailing from the right engine. They reported seeing the airplane bank to the left and a pilot-rated witness reported, '...the right engine was trailing smoke. The aircraft made a turn to the left as if trying to return to the airport. Possibly due to the south tower, he continued to tighten his turn to the left. I told my wife, "he's in trouble." He appeared to stall and went below the tree line.' Several witnesses reported seeing the airplane pitch nose down, impact trees, then the ground, followed by an explosion. One witness reported '...plane's engines were at high speed, no smoke, no fire,' while another witness reported seeing the plane fly over the ocean and '...it sounded like the engine was cutting out....'

"According to the line service employee who witnessed the takeoff, he helped load bags into the airplane and he sat briefly inside the airplane. With respect to the baggage, he noted eight bags of luggage and four sets of golf clubs. One of the bags that he loaded into the aft baggage compartment weighed an estimated 100 pounds. The pilot loaded the golf clubs into the airplane. He thought that each person had two pieces of luggage consisting of a garment bag, and one additional bag that could carry 2–3 days of clothes. Weight calculations were performed using the last known empty weight of the airplane. The airplane was calculated to weigh 5555 pounds at the time of departure. The airplane type certificate data sheet lists the maximum takeoff weight as 5500 pounds."

The National Transportation Safety Board determines the probable cause(s) of this accident as follows:

"The pilot's failure to maintain airspeed (Vs) during a single engine approach resulting in an inadvertent stall. Factors contributing to the accident were a fatigue crack in the exhaust pipe in the right engine, the aircraft weight and balance was exceeded, degraded aircraft performance and the pilot's diverted attention."

Analysis

The errors that led to this mishap can be summed up in one word—*preparation*. Although the aircraft was slightly overloaded and experienced an engine malfunction on takeoff roll, both engines were still producing power at impact and the aircraft was in a flyable condition. The bottom line here is that the pilot simply got overwhelmed by the situation, and it all began when he failed to accomplish a run-up check that might well have identified the engine malfunction before it became an airborne problem that could overwhelm him. Once airborne, the call from the tower and the mechanical problem, when coupled with the workload of having to mentally switch from "takeoff" to "landing" mode, were too much for the pilot to handle, and he forgot rule one—fly the airplane. *Bottom line:* If you are truly ready

for takeoff, make certain you have also adequately prepared yourself for an emergency return to the field.

Some Final Words on Aircraft Performance during Approach and Landing

The famous rhetorical question, "What is most useless to
a pilot?" has many answers (some unprintable), but
most I have heard include "altitude above you" and
"runway behind you." To these we might add, "airspeed
lost" or in some cases "airspeed gained," depending on
the situation. Piloting is filled with decisions, and when
we are talking about issues of aircraft performance,
these decisions are often "one-way decision gates" that
can put you and your aircraft—quite literally—on the
wrong side of the proverbial power curve. If we hold to
the premise that an error chain builds over time, then
we must realize that failures of learning, failures of
planning, and failures of briefing set the stage for a bad
day. Once in flight, actions—such as flight into icing
conditions—that can severely impair performance must
be carefully considered, using a "worst-case scenario"
risk management technique.

Remember the basics of your aerodynamics—lift,
thrust, weight, drag—and keep in mind that your aircraft was designed and built by the imperfect hand of
man. Finally, remind yourself that you may be a great
pilot, but you can have a bad day. Consciously think
through each situation and leave yourself an out.

Let's now turn to the challenges that present themselves when flying into large and small airfields, as we
seek to better understand common pilot errors in these
environments.

6

The Big and Small of It: Operations at Large Airports and Nontowered Fields

Freedom of movement lies in the soul of every aviator. To be able to pick up and go, out of the concrete jungle and carbon monoxide prison of the less fortunate earth dwellers is one of our greatest joys in life. However, unless we fly helicopters or are bush pilots, most of our flight operations begin and end in a structured setting, some of which are concrete jungles in their own right. Pilots' ability to operate across the spectrum of airport complexity is a clear indicator of their professionalism. Yet every year, scores of aviators—and not just from general aviation—get in over their heads at large airports or at small nontowered fields. In this chapter, we will look at some of the common pilot error patterns at each end of the spectrum, and offer some ideas about how to avoid or correct them. Let's begin with a look at the sheer scope of the issue and a discussion of future challenges.

The Challenge of Keeping Up with Aviation Growth

By 2003, it is predicted that the United States will have over 20,000 airports (this figure includes all types of airfields, including military, civil, joint-use, heliports, seaports, and STOL ports), and each and every one of them present the pilot with unique concerns. Even in the aftermath of the September 11 attacks, aviation will grow, although it will take a while for commercial aviation to recover completely. TABLE 6-1 illustrates the magnitude of that growth and sets the stage for our discussion.

6-1 *Number of U.S. Airports**

	1980	1985	1990	1991
Public use:				
% with lighted runways	66.2†	68.1	71.4†	71.9
% with paved runways	72.3†	66.7	70.7	71.5
Total	4,814	5,858	5,589	5,551
Private use:				
% with lighted runways	15.2†	9.1	7.0	6.8
% with paved runways	13.3†	17.4	31.5	32.0
Total	10,347	10,461	11,901	12,030
Total airports	15,161	16,319	17,490	17,581
Certificated:‡				
Civil	§	§	§	§
Military	§	§	§	§
Total	730	700	680	669
General aviation:	14,431	15,619	16,810	16,912
Total airports	15,161	16,319	17,490	17,581

*Includes civil and joint-use civil-military airports, heliports, STOLports, and seaplane bases in the United States and its territories.
†Revised.
§Data do not exist.

With so many potential destinations, a standardized methodology for handling common problems becomes critical for safe operations. Working from large to small, let's tackle this challenge.

Where Are the Nation's Largest Airports?

If you frequently operate out of a large airport, you already know it, as well as its unique challenges. The following is a list of the 50 largest airports in the United States, determined by passenger loads from

1992	1993	1994	1995	1996	1997	1998
72.3	72.8	73.5	74.3	74.5	74.6	74.8
71.6	72.2	72.9	73.3	73.7	74.0	74.2
5,545	5,538	5,474	5,415	5,389	5,357	5,352
6.6	6.3	6.2	6.4	6.4	6.4	6.3
32.2	32.7	33.0	33.0	32.9	33.0	33.2
12,301	12,779	12,869	12,809	12,903	12,988	13,418
17,846	18,317	18,343	18,224	18,292	18,345	18,770
§	§	577	572	577	566	566
§	§	95	95	94	94	94
664	670	672	667	671	660	660
17,182	17,637†	17,671	17,557	17,621	17,685	18,110
17,846	18,317	18,343	18,224	18,292	18,345	18,770

‡Certificated airports serve air-carrier operations with aircraft seating more than 30 passengers.

Source: U.S. Department of Transportation, Federal Aviation Administration, *Administrator's Fact Book,* annual issues.

U.S. Department of Transportation statistics. Some—like O'Hare and DFW—are obvious. Others may surprise you.

1. Chicago (O'Hare), Illinois
2. Atlanta, Georgia
3. Dallas–Fort Worth (Regional), Texas
4. Los Angeles, California
5. San Francisco, California
6. Denver, Colorado
7. Phoenix, Arizona
8. Detroit, Michigan
9. St. Louis, Missouri
10. Las Vegas, Nevada
11. Miami, Florida
12. Newark, New Jersey
13. Minneapolis/St. Paul, Minnesota
14. Houston (Intercontinental), Texas
15. Seattle-Tacoma, Washington
16. Boston, Massachusetts
17. New York (La Guardia), New York
18. Charlotte, North Carolina
19. New York (John F. Kennedy), New York
20. Pittsburgh, Pennsylvania
21. Orlando, Florida
22. Honolulu, Hawaii
23. Salt Lake City, Utah
24. Philadelphia, Pennsylvania
25. Washington (National), D.C.

26. San Diego, California

27. Cincinnati, Ohio

28. Baltimore, Maryland

29. Portland, Oregon

30. Tampa, Florida

31. Oakland, California

32. Washington (Dulles International), D.C.

33. Kansas City, Missouri

34. San Juan, Puerto Rico

35. San Jose, California

36. Ft. Lauderdale, Florida

37. Chicago (Midway), Illinois

38. New Orleans, Louisiana

39. Houston (William P. Hobby), Texas

40. Memphis, Tennessee

41. Orange County, California

42. Dallas (Love Field), Texas

43. Sacramento, California

44. Ontario, California

45. Nashville, Tennessee

46. Albuquerque, New Mexico

47. Indianapolis, Indiana

48. San Antonio, Texas

49. Raleigh/Durham, North Carolina

50. Columbus, Ohio

Obviously, these are not the only airports where pilots experience challenges with complex ground operations or high-density traffic congestion. But if your destination is one of these fields, or even if your flight could conceivably result in a stopover for maintenance or be listed as a weather alternate, you should be ready to operate safely and confidently in the large airport environment. Let's take a look at three of the most common errors associated with operations at large metropolitan airports: airport and runway confusion, communications challenges, and finally, the dreaded "land and hold short" clearance.

Airport and Runway Confusion

One of the most serious issues when operating in a large airport environment is simply the presence of other airports—often large airports—nearby, which can only add to the confusion, as this ASRS reporter discovered in the nick of time.

Case Study: Too Much Trust, Not Enough Procedure (ASRS Accession Number 154320)

"This sequence of trips had me paired with a very experienced, very capable, and very conscientious FAA designated check airman. Owing to his high diligence and capability displayed throughout the course of the trip series, I felt very secure flying with this individual. The final leg of the series ended with the CIVIT two profile descent to a landing on Runway 25L at LAX. Upon being cleared for the profile descent by a LA controller, the Captain/instructor directed me to dial in the Los Angeles ILS frequency used in this descent which he had tuned in his radio. I looked over at what he had tuned in his receiver and tuned the same frequency in mine, 109.7, without consulting my chart. I noticed we were high on the glide slope, with the localizer pegged to the R. I pulled the speed brakes, increasing speed and rate of descent in order to try to cross CIVIT as close to 14,000 ft as possible. At the same time, I began correcting to the R in order to intercept the localizer. I crossed what I assumed to be CIVIT high at FL180 and continued down in order to cross ARNES at as close to 10,000/250 knots as possible. I believe I was very close to 10,000 ft, still no movement on the localizer, when an urgent call from Controller directed me to 'Climb immediately to 14,000 feet.'

"Without hesitation I did this. It was shortly after this that we discovered that the ILS frequency which we had tuned was for Ontario, Runway 26L, about 40 miles closer to us than LA. It is hard to point to determine exactly where we were or how close to the mountains we were. I consider this to be a major altitude bust. My assessment, for myself, of this incident is as follows: I am a professional, and this is a basic violation of procedure. I continue to learn not to trust/assume that anyone's performance will be perfect and it seems always that I relearn this lesson when I am flying with those in our profession who seem the most competent. A simple glance at the chart would have told me that 109.9 was the correct frequency. But I assumed that 109.7 was correct because it came from this individual who had performed so flawlessly up to this point."

Preparation steps that might have avoided this mishap

Clearly, this was a procedural oversight on the part of both pilots, but one that had its roots in preflight planning and briefing. Two aspects of this particular situation trouble me. First, the casual method in which the two highly experienced pilots were operating could have been the result of a missed opportunity to brief how to manage critical changes in communications and flight path. Second, there was breakdown in basic CRM ("a simple glance at the chart would have told me…"), which tells a pilot to use all available resources, including the simple ones, like a chart. This has the potential to underlie a larger problem with personal flight discipline.

In-flight actions

Complacency on the part of both pilots was the major culprit here. Verification procedures exist to ensure that there are no single point failures that can result in a mishap. If this crew would have used the "two-person" concept that is commonplace when changing, for example, automated waypoints, this mistake would have been caught before it was acted upon.

Lessons learned

Aviators need to develop a layered approach where critical information is concerned, checking and validating key information and testing assumptions. Much of this can be preaccomplished during flight planning, some of it can be handled with standard operating procedures, such as the check and verify approach suggested above. In this case, the pilots (and the rest of us) have learned a valuable lesson, and hopefully will not repeat it.

Let's now move toward one of the most common pilot errors at large airfields, where pilots with the best of intentions commit dangerous errors due to communications confusion.

Communications Challenges at Large Airports

Nothing angers me more than a fast-talking approach controller on a windy, rainy night at a large airport. A few years ago, I would have to honestly say that it intimidated me on occasion. It wasn't until I was invited to an approach control facility in Dallas–Fort Worth to watch these men and women in action, that I really understood how much of the responsibility for effective communications rested with *me,* the pilot. Fast-talking controllers still make me mad, but at least now I understand their side of the equation.

There are many causes for communications errors, but at and around large airports, a single communications error has a ripple effect, causing additional transmissions and diversion of critical attention. In addition, unlike many other types of pilot error where it typically takes three to five mistakes to form an accident chain, a *single* misunderstood clearance in a high-density traffic area can have dramatic results. Consider the following event reconstructed from two ASRS reports.

Similar call signs (ASRS Accession Numbers 266870 and 266985)

"We were operating about one hour late, which put us in XYZ area at the same time as [company flight] 552. Our number was 522. Controller cleared 522 direct, descend and maintain 4000.... I acknowledged and we complied. We had not heard 552 on frequency yet. Nor had we heard him respond to the same clearance. We had blocked each other and not known it until a phone conversation later. Suddenly the Controller said, '552, Where are you going?' 552 [replied], 'You cleared us direct down to 4000.' ATC was silent for about 10 seconds, seemed longer.... A target showed on TCAS at 12:00 o'clock, 2000 feet below us. If [we had] continued we would have had a near midair. ATC continued to give both 522 and 552 a lot of strange vectors—obviously for traffic. I queried ATC about it and he said, 'You guys keep getting your flight numbers mixed up.' I know he said 522 in the original clearance, but he meant it for 552. Also 552 was expecting that clearance, so he responded. In retrospect, it was strange that we would be cleared from 9000 to 4000 in such a high density area. I thought maybe the traffic was light at that time."

The Captain of Flight 522 adds:

"No matter how it happened, this is a classic illustration of how dangerous similar callsigns can be, and how a very simple slip by a pilot or controller could result in disaster. My personal feeling is that, given the number of similar callsigns that I hear, my company does not work very hard at 'de-conflicting' them.... The current efforts still leave many problems out there looking for the worst possible time to happen." Scary stuff indeed and it happens all too frequently.

Perhaps the best synopsis and recommendations on the communication issue is contained in Bill Monan's 1991 *ASRS Directline* article titled "Readback, Hearback." I have excerpted portions of the article below and gratefully acknowledge Bill and all the folks at the ASRS for this contribution, which resulted from the analysis of a "data set" of reports from the ASRS system.

Causes of communications breakdown

Why aren't pilots "getting it straight"? We examined a sample set of ASRS reports from airmen and controllers, and identified four major patterns of causal sources for pilot errors in their readbacks.

Readback problems

1. *Similar aircraft call signs.* Airlines, with their hub operations, have set a major trap for their air personnel. Trips 401, 402, 403...Flight ABC1 and XYZ1, GYC and GYE—all operating on the same

frequency, at the same time and in the same airspace. "Good for marketing," protested a reporter, "no good for us."

2. *Only one pilot listening on ATC frequency.* "Picking up the ATIS" and "talking to the company" represented a time-critical gap in backup monitoring during two-pilot operations.

3. *Slips of mind and tongue.* The typical human errors in this category included: Being advised of traffic at another flight level and accepting the information as clearance to that flight level; the classic "one zero" and "one one thousand" mix-up; the L/R confusion in parallel runways; the interpretation of "maintain two five zero" as an altitude rather than an airspeed limitation.

4. *Mind-set, preprogrammed for..., and expectancy factors.* The airmen who request "higher" or "lower" tend to be spring-loaded to "hear what we wanted to hear" upon receipt of a blurred call sign transmission.

The incident set (of ASRS reports) included traffic conflicts, altitude busts, crossing restrictions not made, heading/track deviations, active runway transgressions, and mix-ups of takeoff clearances and parallel runways. Two reports of controlled flight toward terrain were reported.

Hearback problems

"Why didn't the controller catch the pilot error?" was the questioning theme in the data set. While the sources for pilot readback failures were clearly delineated in the narratives, hearback deficiencies diffused into a tangle of erratic, randomly overlapping causal circumstances. But the underlying problem seems to be the sheer volume of traffic: the 9 A.M. to 5 P.M. rush of departures/arrivals; the behind-the-scenes tasks of landlines, phones, and hand-offs; the congested frequencies with "stepped on" trans-missions; the working of several discrete frequencies; and, at times, the time and attention-consuming repeats of call-ups or clearances to individual aircraft. These activities, together with human fallibilities of inexperi-ence, distractions, and fatigue set the stage for hearback failures. Indeed, a series of pilot narratives recognized controller "overload," "working too many aircraft," "over-work," and frequency saturation. These facility conditions provide strong motivations for airmen to drop any "how-the-system-is-supposed-to-work" idealism and adopt a more realistic approach to cockpit communication prac-tices. As a working premise, airmen should assume that during congested traffic conditions, the controller may be unable to hear, or is not listening to, their readbacks.

The ASRS report set included a number of aggressively optimistic assumptions on the part of pilots regarding ATC performance. Reluctantly, but more and more fre-quently, *airmen are accepting silence as a confirmation that readbacks are correct.* Pilots respond to doubtful or partially heard clearances with perfunctory readbacks expecting controllers to catch any and all errors.

Airmen hold to the illusion that ATC radar controllers are continuously observing their aircraft as they progress through the airway structure. The reality is that controllers continually scan the entire scope; they gen-erally do not focus on individual targets. Descent clear-ances that "seem a little early" or altitudes that "seem too low" or turns in the wrong direction may well be intended for another aircraft.

Finally, airmen who fail to brief minimum safe altitudes within or near a terminal area or during the approach phase are vulnerable to readback/hearback errors leading to "controlled flight toward terrain." Such an event is described in an ASRS report from a shaken pilot who admitted to not checking the charts prior to a nighttime descent: "The dim shape of the mountain came into view…seconds before the 'WHOOP…WHOOP…PULL UP' sounded. We both pulled back abruptly on the controls and climbed."

The ATC controller's report added further details: "The tapes revealed that I had told the pilot to descend to 7000 ft (6500 is the MEA) but he had read back 5000. He got down to 5700 ft, about 2 mi from a 5687-ft mountain before I saw him." To summarize the airman: "I don't know how much we missed by, but it certainly emphasizes the importance of good communications between controller and the pilots."

Summary and recommendations on the readback/hearback challenge

When pilots read back ATC clearances, they are asking a question: "Did we get it right?" Unfortunately, ASRS reports reveal that ATC is not always listening. Contrary to many pilots' assumptions, controller silence is not confirmation of a readback's correctness, especially during peak traffic periods.

Pilots can take several precautions to reduce the likelihood of readback/hearback failures:

- Ask for verification of any ATC instruction about which there is doubt. Don't read back a "best guess" at a clearance, expecting ATC to catch any mistakes.

- Be aware that being off ATC frequency while picking up the ATIS or while talking to the company is a potential communications trap for a two-person crew.

- Use standard communications procedures in reading back clearances. "Okays," "Rogers," and mike clicks are poor substitutes for readbacks.

- Controllers can also take steps to safeguard against readback/hearback failures.

- Be aware that an altitude mentioned for purposes other than a clearance, such as a traffic pointout, may occasionally be interpreted by pilots as an instruction to go to that altitude.

- Deliver cautionary messages such as "similar call signs on frequency" to help reduce call sign confusion.

The consequences of readback/hearback failures vary, but when they occur in the context of high rate of climb/descent operations, ASRS reports frequently conclude: "It was too late to intervene—the aircraft had already passed through an occupied altitude."

The future

Reflecting a major trend in ASRS data, the report set poses troublesome questions concerning the ATC-pilot communications procedures. Are traffic growth and congested frequencies compressing the traditional to-from-to exchanges into a one-way transmission? Are airline managements aware of the similar call sign problem? Are airmen placing full-time confidence upon a confirmation procedure that works only part of the time? Can data link help solve some of these problems? Postulated a pilot reporter: "If, in truth, controllers are unable to listen, then we should change the system."

Bill Monan's observations and questions are as valid 10 years after this article was written as they were then. Although flight training programs now emphasize communications and standard phraseology, we still have a long way to go to completely conquer human communications error in the large airport environment. This is true even after we have successfully made an approach and landing.

The third big challenge in the large airport environment comes after we touch down, a time when some pilots drop their vigilance guard. This can result in serious embarrassment, and occasionally much worse.

The Dreaded "Land and Hold Short" Clearance

One of the most hazardous errors a pilot can make is failing to follow "hold short" instructions following a landing. Often, our mind relaxes a little bit after we touch down and slow to taxi speed, but this can be a deadly mistake. Consider the close call with the following ASRS reporter:

"My First Officer was flying. A military transport had landed on Runway 24[R], [and] was instructed to turn left and hold short of 24L.... We were cleared for takeoff on Runway 24L.... Just prior to lift-off speed, we observed [the] military transport start taxiing. He taxied onto runway 24L in front of us. We were then above the V1 speed, and our only option was to continue the takeoff. We were able to lift off over the military transport, but had our gross weight been closer to maximum, we might have had a real problem. [Upon] arrival at our destination, I called the [departure airport] Tower, and their people confirmed that the military transport had crossed Runway 24L without a clearance."

This was certainly an incident that might well have resulted in multiple fatalities. We do not know why this particular crew failed to comply with the instructions, but we do know that most often unfamiliarity with the airfield is a key component of runway incursions. We have dedicated an entire book in the Controlling Pilot Error series to runway incursions (*Runway Incursions* by Bill Clarke), so we won't go into great detail here, except to repeat that famous aviation saying that tells the pilot to "fly the aircraft until the last piece stops moving" and that includes following all taxi instructions to the chocks. At the opposite end of the spectrum, at small nontowered airfields pilot error also presents a serious safety challenge. Even though the challenges are different, human failures often result in the same tragic ending.

Nontowered Airports

Many pilots, myself included, feel naked when we leave the comfort zone of controlled airspace. The idea of having a third party responsible for assisting in separating my aircraft from *terra firma* and other pieces of flying metal is warmly reassuring. The rapid-fire chatter of a big city air traffic controller is music to my ears, and the relative silence of VFR operations in remote settings to me is vaguely reminiscent of an old horror movie where the boogeyman is lurking out there somewhere, waiting to pounce.

I know, however, that there are others who feel just the opposite, and are most at home on a VFR flight plan flying into a rural landing strip, broadcasting their intentions in the blind on CTAF frequencies. In spite of the myriad of differences, the goal of both operations is identical, safe aircraft separation. The essence of the difference can be captured in the phrase "controlling influence." In the large airport environment, the controlling influence is obviously provided by the ATC. At the nontowered airport, it is provided by the *participating* pilots. The problems arise from the fact that not all pilots choose to participate.

Although the goals are the same, there is an important—and misunderstood—distinction between the two environments. Under instrument flight rules, compliance with ATC directives is mandatory. At nontowered airports, standard practices are recommended. Herein lies one of the most critical concerns regarding common pilot errors. At the end of the day, the responsibility for safe operation in either environment lies with the pilot. That is the law as written in the Federal Aviation Regulations, the rest are just tactics to achieve the objective of safe operations.

In a recent issue of *FAA Aviation News,* pilot and FAA Program Manager Patricia Mattison tells of a close call she had at a nontowered airport, caused by a pilot who confused "legal" with "safe":

"Several years ago, when I had a Cessna Pilot Center, I was out doing touch and go landings with a relatively new student pilot.... The cloud bases were at 1500 feet and aircraft had been making the VOR approach, circle to land, at the field for most of the day. My student was making all of the appropriate position reports on downwind, base and final, as well as announcing that we were going to make touch and go landings. At one point we had turned to final approach and were well established on the approach when I heard, from the ground, 'Patti make an immediate left diving turn—NOW!!!' Recognizing the voice and urgency, I instinctively complied. A twin Cessna that had been on a long straight-in approach flew *within a few feet* of us, narrowly missing my plane, and landed. Had I not automatically reacted we would have been involved in a mid-air collision with the twin.

"A quick thinking former student on the ground in the run-up area had the presence of mind to alert us of the impending danger. As a result of that warning my student and I avoided becoming a statistic. Later, after I quit shaking—not out of fear but with fury—I went to the pilot of the twin. I read him the riot act, loud and long, about announcing position and listening on the radio for traffic in the pattern. Comment from the pilot was that it was legal to land from that approach. Legal it might have been, but not a safe thing to do at a non-towered field."

Patricia's quick response in the air was mirrored by her good judgment on the ground. Confronting a fellow pilot is never an easy thing to do, but it is always the *right* thing to do, if it is done in the spirit and in the name of safety. The offending pilot should have offered his thanks, not only for the avoidance maneuver, but also for the words of wisdom.

Now that we have addressed the question of legality, let's dive right into common errors committed by pilots operating in this less structured environment.

The Three Deadly Sins

I was once told by a wise flight instructor that any flight where you make only "new mistakes" is an opportunity to measure your professional growth. His point was simply that we should not need to make the same mistake twice, and every new error is fertile ground for improvement. We must also learn from the mistakes of others. The following three errors—inappropriate pattern entry, failure to make pattern position reports at the appropriate times, and impatience—are the most frequent and lethal errors made by pilots operating at non-towered fields. Refer to FIG. 6-1 during the discussion to orient yourself with pattern entry procedures and reporting points.

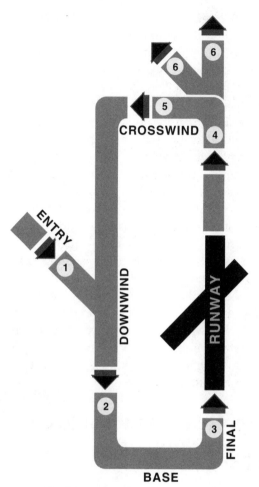

6-1 *Pattern entry and reporting points.*

Inappropriate Pattern Entry

Entering the traffic pattern at a nontowered airfield is an art. It begins with preflight planning, where you take the time to familiarize yourself with the local frequencies and procedures. We will begin by assuming that as a licensed pilot you are familiar with the applicable portions of the FARs, contained in Parts 91, 93, and 97 and consolidated in some excellent advisory circulars from the FAA. A serious study of the Airport/Facilities Directory (AFD) is in order, as is a review of any NOTAMs.

Armed with the information, your entry begins 10 to 15 mi from the field with the initial call-up, typically on UNICOM frequency, providing your aircraft identification, location, altitude, and intentions. You should also request traffic and any other airfield advisories. Finish all calls with the name of the airport, so as not to confuse others operating within radio range at other airfields.

The preferred arrival begins with an overflight of the field, descending away from traffic using clearing turns (see and be seen) and entering at a 45-degree angle to the downwind. To be certain, there are times and situations when other entries make sense, but be advised that many conflicts and midair collisions could be traced to one simple decision. "How am I going to enter the pattern?" If you are having second thoughts, you should ask yourself, "Is my decision based on safety or convenience?"

Case Study: Heard But Not Seen (NTSB Report Number LAX96FA108A)

"On February 8, 1996, at 1052 hours Pacific standard time, a Piper PA28-181, N791CR, and an Avions Pierre Robin R2160, N216JM, collided about 3 miles southwest of Ramona Airport, Ramona, California. The pilot in each airplane was conducting a visual flight rules (VFR) personal flight to Ramona Airport. N791CR landed on runway 27 at Ramona Airport and N216JM crashed and burned in an open field about 1 ½ miles southwest of

Ramona Airport. The Piper sustained substantial damage; the Avions Pierre Robin was destroyed by impact and the post-impact fire. Visual meteorological conditions prevailed.

"National Transportation Safety Board investigators interviewed the Piper pilot at Ramona Airport. The pilot said that he was flying to Ramona Airport to practice some touch-and-go landings and takeoffs. The pilot flew to the practice area after departing Palomar Airport, and the pilot executed a few S turns after he arrived at the practice area and proceeded toward Ramona Airport.

"When the flight was about 2 miles east of Mt. Woodson he heard some airplanes on Ramona Airport's common traffic advisory frequency (CTAF; 122.7 MHz). He said that runway 27 was in use and he continued toward Mt. Woodson to enter the left traffic pattern at a 45-degree angle. The airplane's altitude was 2700 feet mean sea level (all altitudes in this report, unless otherwise noted, are mean sea level altitudes).

"When the airplane was abeam Mt. Woodson he began a descent and announced his position and landing intentions on runway 27 on the CTAF. Moments after leaving 2000 feet, he saw a blue and white flash off and below his right wing and then felt the impact. Other than the blue and white flash, he never saw N216JM. After the collision, the pilot declared an emergency and landed on runway 27. He said that he did not realize that the airplane's right main landing gear separated at the collision. On touchdown, the airplane began to skid to the right and then stopped. The pilot exited the airplane without any assistance.

"Safety Board investigators interviewed the Avions pilot at Palomar Hospital, Escondido, California. The pilot said that he was going to Ramona Airport to drop off some keys to his mechanic. He said that he received a weather briefing from the FAA, San Diego Flight Service Station, before departing on the accident flight. He said that he received VFR radar advisories from the FAA, South Coast TRACON (Terminal Radar Approach Control). When the flight was a few miles southwest of Mt. Woodson, the sector controller terminated the radar advisory service.

"When the airplane was abeam Mt. Woodson, descending through 2600 feet, he heard another airplane report in the vicinity of Mt. Woodson. He looked for the airplane, but did not see it. Moments later, the canopy blew off and the engine began to vibrate severely. He elected to land in an open field. The airplane bounced when it touched down and almost immediately erupted into flames. The pilot exited the airplane when it came to rest.

"In a follow-up telephone interview conducted on February 9, 1996, the pilot said that the airplane was level at the time of the collision. He said that the airplane's altitude was 3500 feet when the sector controller terminated the radar services and he descended the airplane to 2400 feet. While level at 2400 feet, he heard the other airplane announce his position moments before the collision.

"A ground witness that saw the airplanes before and during the collision reported that he was in front of his residence when he saw the airplanes. He said that both airplanes appeared to be flying in a formation with one airplane above the other. He became distracted momentarily and then returned his attention to the airplanes. The lower airplane [N216JM] then appeared to climb into the other airplane. The upper airplane [N791CR] continued its flight to the airport. The lower airplane [N216JM] crashed in an open field."

So what went on here? Before reaching the traffic pattern, both pilots heard other traffic on the common traffic advisory frequency (CTAF). As they were positioning themselves to enter a left traffic pattern at a 45-degree angle, a midair collision occurred.

The National Transportation Safety Board determined the probable cause(s) of this accident as follows: "Inadequate visual lookout by the pilots of both airplanes, which resulted in their failure to see and avoid each other."

But there was a bit more here than that, wasn't there? Neither pilot made the recommended call-up to request traffic advisories. The Piper pilot made a descending entry assuming his flight path was clear. The Avions Pierre Robin pilot was flying directly beneath the Piper for some period, and even though he heard a call from an aircraft "abeam Mt. Woodson" he continued without visual contact. If he could not see the other aircraft he assumed that there was not a conflict. A more conservative approach is to assume that the aircraft is there, in a blind spot. Turn your aircraft to a known clear area, search again in areas that you could not see before—and reenter when you *are certain* that the flight path is clear. Both pilots were extremely fortunate to survive, and you can bet that they have learned from their mistakes—we can too.

In the case above, both aviators made the recommended 45-degree to downwind entry, yet it was flawed just enough to result in a midair. But all the mistakes could have been mitigated with effective communication, the next topic of discussion.

Traffic Pattern Position Reports

Perhaps the most frequent error pilots make is failing to monitor the CTAF frequency and make the initial call-up early enough, followed closely by a failure to make all recommended calls as suggested by the AIM. FIGURE 6-1 illustrates locations where recommended traffic pattern position reports should be made. This communications discipline is key to successful operations, but keep in mind that radios are not required for operations at non-towered airports, and even when they are installed, the pilots may not be listening, or as in our next case study, the equipment may fail at the worst possible time.

Case Study: What We Have Here Is a Failure to Communicate (NTSB Report Number MIA96LA120A)

"On April 16, 1996, about 1925 eastern daylight time, a privately owned Mooney M20C, N9206V, and a privately owned homebuilt Christian Eagle, N83BL, collided while both were on final approach to land at the St. Augustine Airport, St. Augustine, Florida. Visual meteorological conditions prevailed at the time and no flight plan was filed for either flight, which were both operating as a 14 CFR Part 91 personal flight. The Mooney sustained minor damage and the private-rated pilot and one passenger were not injured. The Eagle sustained substantial damage and the private-rated pilot, the sole occupant, was not injured. The Mooney flight originated about 1800 from the St. Lucie County International Airport, Fort Pierce, Florida. The Eagle flight originated about 10 minutes earlier from the St. Augustine Airport.

"The Eagle pilot stated he had performed one touch-and-go landing and the flight remained in the traffic pattern using left hand turns for a planned full stop landing. While on final approach to runway 31 about 150 feet above ground level, he felt an impact and observed damage to the right wing of his airplane. He landed with no further damage to his airplane. Post accident airplane damage assessment indicates in part that the right main landing gear tire was contacted by the propeller of the Mooney. Also, propeller slash marks were located on the bottom of the fuselage from the wings to the empennage. An individual who was monitoring the UNICOM radio and another pilot who was inbound to the airport both stated that they did not hear the Eagle pilot announce his landing intentions.

"The Mooney pilot stated that while near the airport, he monitored the UNICOM frequency and entered the traffic pattern to land using left hand pattern, for runway 31. He announced all legs over the UNICOM frequency and while on short final approach about 100 feet above ground level, he felt a jolt and the airplane was forced down. He recovered and landed with no further damage to his airplane. Post accident airplane damage assessment indicates that the propeller was damaged. The individual who was monitoring the UNICOM, and the pilot of another airplane who was inbound to land both reported hearing the Mooney pilot announce his landing intentions over the UNICOM frequency. The pilot of the other airplane also reported seeing the Mooney in left traffic for runway 31.

"The Christian Eagle pilot said that after completing a touch-and-go on runway 31, he remained in a standard left pattern. He stated that he had been listening to other aircraft and reporting his position in the pattern, but on downwind after the touch-and-go, his 'radio started to break up on transmission.' He stated that he turned base and final calling his position; however, neither the UNI-COM operator nor pilots of other airplanes heard the Christian Eagle pilot announce his landing intentions."

The National Transportation Safety Board determined the probable cause(s) of this accident as follows: "Inadequate visual lookout by the pilots of both airplanes, which resulted in their failure to see-and-avoid each other's airplane. A factor relating to the accident was: the lack of traffic pattern position reports from the

Christian Eagle, due to the reported failure of its VHF radio for undetermined reasons."

Lessons learned: When things change...

This (Eagle) pilot knew, or at least suspected, that he was having radio problems, yet took no apparent action to mitigate the new risk associated with the developing equipment failure. This is a valuable lesson for all of us. When things change—be that weather, traffic pattern direction, aircraft equipment, or any other significant factor associated with a risk decision, we need to reevaluate our previous decisions and intent on the basis of the new situation.

Impatience

Pilots often succumb to temptations to deviate from recommended practices simply because they are in a hurry. Over time, these "just once" deviations become common practices. Transport Canada published an Aviation Safety Letter on this subject titled "Dangerous Practices Becoming Common at Uncontrolled Aerodromes," reprinted in part below:

"Few small aerodromes benefit from the luxury of parallel taxiways or holding bays near the runway threshold. They are one-runway operations. Arriving and departing aircraft have to sequence themselves properly to avoid conflict. It can be particularly annoying when the parking area is at the far end (of the runway) and a long taxi is involved before a pilot can get into position to safely do a run-up and depart. Some pilots have to wait to taxi, or others have to wait to land.

"As a result, in the interest of expediting traffic, pilots are developing dangerous habits, habits that are not only being accepted but also, on occasion, *being taught by instructors* [emphasis added]. Pilots create their own parallel taxiways in the grass, just off the runway. These are being used while other aircraft are arriving and departing. Aerodrome standards require that parallel taxiways be far enough from the runway to guarantee wingtip clearance plus a big safety margin. This means several hundred feet away, not just off the runway surface on the nicely graded and prepared area.... By mutual arrangement some pilots are landing over the top of other aircraft, some are backtracking, and others are waiting for takeoff on the threshold.

"Last, but not least, parallel takeoff/landing operations occur with some pilots using the runway while others use the adjacent grass; the runway user conform(s) to the recommended left hand circuit and the grass users do both left and right hand circuits.

"These are very dangerous practices. Picture a sunny weekend when everybody wants to fly. Picture a couple of arriving and departing transient pilots who don't know the local habits. Picture a collision.

"Sometimes you just have to wait your turn."

This letter speaks for itself. As action-oriented people, pilots can be their own worst enemy. In the structured airspace around a large airport, this impatience may result in frustration evidenced by a curt radio call, but the pilots are forced to wait their turn by air traffic controllers. When the pilots themselves become the controlling influence, this impatience can result in hazardous attitudes and unsafe shortcuts.

Keys to Overcoming Common Errors

According to an ASRS *Callback* publication on the subject, there are several keys to combating common pilot errors at airports that do not have an operating control tower. One is selection of the correct common traffic advisory frequency (CTAF), the frequency designated for carrying out airport advisory practices. The CTAF may be a UNICOM, MULTICOM, FSS, or tower frequency and is identified in appropriate aeronautical publications. Another crucial practice is careful visual scanning to see and avoid other aircraft, especially those that are low flying, on straight-in approaches, or have no radios. Recent ASRS reports illustrate several problematic scenarios that can occur at fields without operating control towers. In our first example, two aircraft appear to be playing the old teenage highway game of "chicken."

Opposite Direction on the Same Runway

"We were departing [nontower field] on Runway 05, wind was calm. Made numerous calls that our King Air was taxiing from the ramp to Runway 05, departing Runway 05, etc. [Piper] Archer made no radio calls. We heard other aircraft in pattern using Runway 05. As we were accelerating to V1, the [Piper] Archer came over the hump in the runway. He was using Runway 23. I swerved right and aborted the takeoff. We missed by 30 feet. The fact that the ends of the runway are not visible from one another was a major factor in this occurrence. Also, the fact that the [Piper] Archer made no radio calls is the most important factor. The best way to avoid this problem in the future is for [pilots] to...follow standard established procedures for operating in non-controlled environments."

The obvious lesson here is one of vigilance, as well as preparing oneself to deal with the immediate unexpected event. This pilot left himself an out, and he needed it. The root of this error may be traceable to a common pilot tendency at nontowered airports illustrated in our next case study.

Case Study: Use of a Nonpreferred Runway

"We reported on CTAF 119.4, B737 pushing back for departure, Runway 14 [airport]. Tower does not open until 0500 local time. At 0445 we reported B737 taxiing for Runway 14 [airport]. We noticed one aircraft in the pattern for Runway 32 that reported downwind for Runway 32. At 0450 we were released by Approach for departure. We verified the aircraft visually on downwind abeam midfield and reported on CTAF, 'B737 departing Runway 14 [airport].' Immediately, we heard on CTAF, 'I'm on my takeoff roll, Runway 32.' At the far end of the runway we notice a Cessna on takeoff roll…. We delayed our takeoff.

"This could have been a safety hazard if the other aircraft had not made the radio call…. Factors affecting the situation: (1) Runway 32 is the preferred runway we were using Runway 14 for our direction of flight; (2) We wanted to depart promptly to avoid a conflict with the downwind aircraft."

"Could have been a safety hazard?" Boy, is that the understatement of the year. This reporter mentioned that he was the only one monitoring CTAF while his first officer was getting the clearance from approach control, and for this reason the crew might have missed the Cessna's earlier CTAF calls. Pilots on IFR flight plans, like this crew, also have the option of asking approach control whether there is traffic inbound for the runway in use. For some pilots there is no option but see and avoid, because of situations like the ones we see in the ASRS report that follows.

Case Study: No-Radio Aircraft

"While taxiing to the end of Runway 03 for takeoff I checked the runway and traffic pattern. I had no radio so I couldn't hear the radio traffic. I was checking the runway for traffic and saw none. As I crossed the hold line for Runway 03 I checked again and saw a Cessna 172 bounce after touchdown on the runway. I immediately stopped halfway between the hold line and the edge of the runway. After the C172 turned off the runway and passed me, I back-taxied to Runway 21 and took off… The Cessna apparently blended in with the background of trees and buildings and I did not see the Cessna until it bounced after touching down."

The Aeronautical Information Manual offers a comprehensive summary of recommended communications procedures at airports without operating control towers, and emphasizes that not all aircraft operating into these fields have radios. This gives us all good reason to keep our eyes out of the cockpit and our head on a swivel.

Summary

This chapter brought out many typical pilot errors that occur in and around large airports, as well as a unique set of error patterns associated with nontowered fields. At the end of the day, the pilot's job is not much different in either environment. We must:

1. Ensure safe entry into an orderly flow of traffic.
2. Comply with the guidance in place.
3. Clearly communicate our intentions.
4. Remain constantly vigilant for other aircraft.
5. Be prepared to handle the unexpected.

Whether we are talking about voluntary *compliance* with FARs or voluntary *observance* of SOPs, a pilot's job is the same in the high-density traffic area of a large metropolitan airport and a rural nontowered field. It's as simple as being where you are expected to be.

In this chapter we saw how pilot errors at large and small airports can result in failed separation, and on occasion a midair collision. Let's now investigate the midair potential more closely, and identify the types of pilot error that put us at greatest risk.

7

Midair Collision
Avoidance

Perhaps nothing in aviation strikes more fear into the hearts of an aviator than the term "midair." It conjures up pictures of sudden aluminum overcast, sounds of crunching metal, loss of control. These are accurate representations of what actually occurs when two aircraft come together in flight. I know, I was in one.

On December 9, 1983, I was serving as a copilot aboard a USAF KC-135 pulling duty in the Persian Gulf region, when we were struck by an E-3 AWACS aircraft following an in-flight refueling. Although the incident was not during the approach and landing phase of flight, it did involve the classic human errors of distraction, communication breakdown, and the perceived need to "do something different." I will never forget the way the huge aircraft suddenly filled the windscreen, or the voice of someone in the cockpit yelling "he's going to hit us." The loud high-pitched squeal of metal on metal is forever etched in my long-term memory, followed by moments of near panic as we fought to regain control of the aircraft. In this case, no one was killed in the mishap. Even though the damage was quite severe (two engines lost on one side, large fuel leak, fire indications, major structural damage to right wing), the Lord decided that it was not our day to die, and we got to write our own addition to the "blood regulations" (rules that emerge from mishaps) prohibiting in-flight amateur photography and "flying two aircraft in close vertical proximity."

I also remember how it all unfolded, one small mis-step at a time. In the nearly two decades since the incident, I have learned that this was a classic "mishap chain" and one that plays out far too often during the approach and landing phase of flight, when a pilot's attention is often channeled on checklists, the runway environment, or an airport diagram, while another fly-ing object closes the distance between aircraft, equally unaware that the "time to die clock" is ticking. I'd love to finish my "there I was" story, but I'll leave that for another time and place because the subject matter of the day is approach and landing error.

Is Technology the Answer?

Technology has long been seen as the answer to midair collisions, beginning as far back as 1956, when a DC-7 and Lockheed Constellation collided over the Grand Canyon killing 128 crew and passengers. This mishap led to the foundational development of our modern air traffic control system. Thirty years later, following a highly publicized mishap between a DC-9 and single-engine Piper over Cerritos, California, Congress stepped in with the Airport and Airway Safety Expansion Act, which requires air carrier aircraft (civil) to be equipped with collision avoidance systems. Collision avoidance technology such as TCAS, TCAD, and ADS-B systems has steadily improved and provides a pilot with another tool to assist in risk management. Notice I said "another tool," because even since the development and implementation of collision avoidance technology in many commercial and general aviation aircraft, the number of midair collisions occurring has remained relatively stable at between 15 to 20 per year. Why is that?

Some speculate (and I count myself among this group) that collision avoidance systems like TCAS have made pilots less vigilant by creating a false sense of security. Recently, while flying with a major airline crew as a jump seat guest, I observed a crew that flew into a large regional airport (with active NOTAMs on glider and parachute operations nearby) and had their heads down in the cockpit for over *14 consecutive minutes* as they descended from 12,000 ft to 1500 AGL on the final approach in visual conditions. After the flight, I asked the captain about the crew's clearing technique and he replied, "Captain, TCAS takes care of that for us and besides, we had you looking." I hated to break the news to him, the last time I checked, parachutes and gliders were not often equipped with transponders.

This attitude is nothing new or limited to aviation. Many safety devices often become used as performance enhancements, reducing the safety aspect and purpose of their design. A classic example of this was illustrated in a study of inner-city cab drivers that were given new antiskid brake systems. Initially, rear-end collisions dropped dramatically in the cabs equipped with the new brakes. However, after a few months, the accident rates returned to their previous levels. Initially, this puzzled researchers, who thought the systems were losing their effectiveness over time. But it was soon discovered that the cab drivers had learned that the new brake system allowed them to drive much faster. Perhaps we are seeing the same phenomenon with the collision avoidance technology. Pilots don't clear as much as they used to because they don't think they need to. Technology has created a false sense of security. Perhaps then, technology, although certainly a part of the solution, may also be a part of a new problem.

The Regulatory Perspective

If technology is indeed only part of the solution, where can we find the rest of the picture? The rules for see and avoid in visual meteorological conditions are contained in FAR 91.113 as follows: "When weather conditions permit, regardless of whether an operation is conducted under instrument flight rules or visual flight rules, vigilance shall be maintained by each person operating an aircraft so as to see and avoid other aircraft."

"See and avoid" sounds so simple, but in reality very few pilots optimize their ability to accomplish this critical component of safe operations. In the paragraphs and pages that follow, we will help you develop the tools and techniques to clear your approach and landing airspace for potentially hazardous traffic and hone your skills for safer operations. Let's begin with a look at one example when things did not go exactly as planned on an otherwise routine day flight for two crews operating out of Fulton County, Georgia.

Case Study: Crossing Paths in the Pattern (NTSB Report Number ATL93FA061A)

"On February 19, 1993, at about 1057 Eastern Standard Time, a Robinson R22B, N621SG, and a Cessna 152, N5532Q, were destroyed following a midair collision over the Fulton County Airport in Atlanta, Georgia. Both aircraft were destroyed in the collision, and the four crew members on board the two aircraft were fatally injured. The Robinson R22B and the Cessna 152 were being operated under 14 CFR Part 91. Visual meteorological conditions existed at the time, and no flight plan had been filed by either aircraft for the local training flights. Both aircraft departed Peachtree Dekalb Airport at about 1000.

"The Robinson R22B helicopter departed Peachtree DeKalb Airport at about 1000 for the purpose of conducting a helicopter instrument rating practical examination. At 1032:30 the pilot of the helicopter, N621SG, contacted the Atlanta Approach Control and advised the controller that the aircraft was being used for the purpose of conducting an instrument check ride, and requested a Very High Frequency Omnirange (VOR) approach to Runway 26 at the Fulton County Airport. The pilot requested that this approach be followed by a missed approach, and then proceed to the Initial Approach Fix (IAF) for a holding turn and an Instrument Landing System (ILS) approach to Runway 8 at the Fulton County Airport, and then to be followed by an Automatic Direction Finder (ADF) approach to Runway 8.

"At 1033:41 Atlanta Approach Control contacted the Fulton County Airport Control Tower and advised the tower personnel that the helicopter intended to execute a VOR approach to Runway 26. The Fulton County Tower approved the request and Approach Control cleared the helicopter for the approach. At 1036:10 the pilot of the helicopter contacted the Fulton County Tower and the tower advised the helicopter that it appeared to be about one half mile north of course. At 1038:23 the control tower cleared the helicopter for the low approach, and again advised that the helicopter was north of course. The pilot advised that they were following the instrument indications in the cockpit and requested a missed approach to the IAF for the ILS. The tower cleared the helicopter for a missed approach direct to the IAF. At 1042:37 the pilot of the helicopter contacted Atlanta Approach, advised that he was executing the missed approach, and was cleared to execute one turn in the holding pattern and then to execute the ILS approach to Runway 8. At 1054:51 the helicopter was advised to again contact the Tower.

"At 1055:03 the pilot of the helicopter contacted the Fulton County Airport Air Traffic Control Tower, and was cleared as number two for the approach following a Piper Seminole, which was pointed out to the helicopter. At 1055:20 the pilot of the helicopter stated 'Okay we got the Seminole in sight six two one sierra golf we'll be low approach back around for the NDB.' The tower personnel responded 'Helicopter one sierra golf roger number two Runway eight cleared low approach.' At 1056:51 Fulton County Tower advised the helicopter 'Helicopter one sierra golf Cessna ahead and to your left touch and go Runway niner.' The pilot of the helicopter responded 'Sierra Golf.' At 1057:40 the pilot of a twin Cessna N3417G asked the tower what the explosion over the field was, and the tower responded 'a helicopter and a Cessna.'

"The Cessna 152, N5532Q departed Peachtree Dekalb Airport at about 1000 for the purpose of flight instruction. At 1046:46 the pilot of N5532Q contacted the Fulton County Air Traffic Control Tower and requested landing instructions for Runway 9 at the Fulton County Airport. The Tower instructed N5532Q to report left downwind for landing Runway 9. At 1051:52 N5532Q reported left downwind for landing Runway 9. The tower cleared N5532Q for touch and go landing on

Runway 9 and advised the pilot of King Air traffic three and one half miles west of the field landing on Runway 8. At 1055:47 the Tower again cleared N5532Q for touch and go landing on Runway 9. No further conversations were made with N5532Q. During the approach to the airport and landing approach of N5532Q, the Tower pointed out numerous traffic to the pilot of N5532Q, however, the Tower did not point out the helicopter, N621SG, to the pilot of N5532Q.

"Witnesses stated that at the time of the midair collision, the Cessna was apparently descending on final approach for Runway 9, and it appeared that the helicopter was executing a climbing left turn just past the approach end of Runway 8. The witnesses stated that the helicopter appeared to overtake the Cessna."

The National Transportation Safety Board determines the probable cause(s) of this accident as follows:

"The failure of the pilot/applicant of the R22B, to follow low approach procedures, and the failure of the designated examiner/observer in the R22B to maintain visual separation from the Cessna 152. Contributing factors to the accident were: the insufficient response of both pilots of the R22B in reply to traffic information, the local controller's failure to obtain positive confirmation of visual acquisition of the Cessna 152 from the pilots of the R22B, and the failure of the local controller to remain attentive to the converging traffic and issue appropriate instructions to avoid the collision."

So what happened here? In a nutshell, the CE-152 was on final approach for a touch-and-go on Runway 9. The R22B was executing a practice ILS instrument approach to Runway 8, had been cleared for a low approach, and had been given an advisory as to the CE-152's location and intention to which the pilot responded "Sierra Golf," apparently acknowledging the callout—but not necessarily confirming that he had the traffic in sight—using nonstandard phraseology. Witnesses stated that the R22B appeared to be executing a climbing left turn just past the approach end of Runway 8, when he overtook and collided with the Cessna.

Preparation steps that might have avoided this mishap

Prior to operating at any airfield, it is prudent to study the airfield diagram and consider where potential conflicts might occur. Additionally, the disciplined use of standard phraseology is something that we need to practice every day. The casual reply to a traffic callout with a call sign "Sierra Golf" is indicative of a personal philosophy of "its always been good enough before." In this case, a single call of "traffic not in sight" could have prepped the controller team to take more aggressive action.

In-flight actions

From purely the regulatory perspective, the pilots did not comply with the intent or letter of the Airman's Information Manual (AIM), which advises "unless otherwise authorized by ATC, the low approach should be made straight ahead, with no turns or climb made until the pilot has made a thorough visual check for other aircraft in the area."

Lessons learned

1. Study airport diagrams to determine likely points of congestion and conflict as a routine part of your preflight preparation.
2. Use standard phraseology as contained in the Airman's Information Manual.

3. Don't rely exclusively on the controller for traffic separation—remember that they are often conducting training too.

Let's now move toward a more fundamental approach to "see and avoid" with a short discussion and example on physiological preparation for midair collision avoidance.

Limitations of the Eye and Scanning Techniques

There is a really cool television advertisement for Pepsi that shows baseball slugger Ken Griffey, Jr., standing at the plate awaiting a pitch. The narrator says "This is what a 96 mile per hour fastball looks like to Ken Griffey, Jr." As the pitcher releases the ball, it slows to superslow motion, each and every seam on the ball clearly visible. In fact, Griffey has so much time available as the pitch approaches, that he spots fellow slugger Sammy Sosa in the dugout trying to sneak a drink of his Pepsi, so he rips the pitch on a line drive into the dugout as a warning shot. Wouldn't it be nice to have that type of vision to spot potential conflicts in the air?

To be certain, professional athletes have developed their visual techniques to an extremely high level. Anyone who has ever watched John Stockton throw a no-look pass or seen Doug Flutie "sense" a blind-side hit and step up in the pocket at the last second knows that these skills are built upon exceptional peripheral vision. But our job as pilots is much more difficult. For one thing, we don't have opponents wearing different-colored uniforms, and for another our adversary can approach from above or below. However, like professional athletes, we can—and must—train our eyes to accomplish our mission objectives.

Obviously, the key to see and avoid is vision. Understanding the strengths and weaknesses of the human eye is as important as understanding your aircraft systems. But many pilots take their vision for granted, and don't take the time to fully develop this marvelous gift of sight. Eighty percent of our sensory input in aviation is through our vision, so let's take a brief look at how to fine-tune this tool.

The foveal field

On the central part of the back of the eye, also known as the *retina,* there is a point where our vision is most acute known as the *fovea.* (See FIG. 7-1.) This small spot is able to process only one degree of vertical and horizontal vision. According to an AOPA *Safety Advisor:*

> "This area of focus is the equivalent of a quarter seen from one eye at a distance of four and one half feet. Anything outside of this area will not be seen in detail. Outside of a 10-degree cone concentric to the foveal cone, visual acuity is only 10 percent of that of the foveal field of vision. In practical terms, a plane that was visible in the foveal from 5000 feet away would only be visible at 500 feet or less if it was more than five degrees on either side of the core vision."

That is quite an evaluation. Think about it for a moment, what is the difference between 5000 ft and 500 ft in a 300-kn head on closure encounter? Probably only the difference between life and death.

So what does this mean for us as pilots? Simply that we need to move our eyes to see what we need to for safe and effective operations, and there are some helpful hints that can assist us in doing so in a systematic manner.

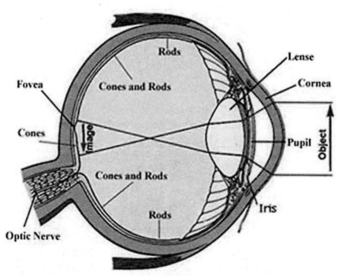

7-1 *The human eye. Note the small fovea, the key to high-acuity vision, but able to process only 1 degree of horizontal and vertical vision, highlighting the need for a highly disciplined scan pattern.*

When and Where to Look

If we knew where a conflict was about to emerge we would certainly direct our attention to that point. Although we may never have 20-20 foresight, we *can* build upon statistics that indicate where midair collisions have occurred in the past, and tailor our see and avoid strategies toward these high-risk areas.

According to statistics from the FAA and NTSB, most midair collisions occur in day VFR conditions, between 10 A.M. and 5 P.M. on weekends, within 5 mi of an airport. Most occur in the traffic pattern and nearly half of these on final approach. Most significantly, *a whopping 82 percent of midair collisions occur at overtaking convergence angles,* and about a third from nearly straight behind another aircraft (see FIG. 7-2).

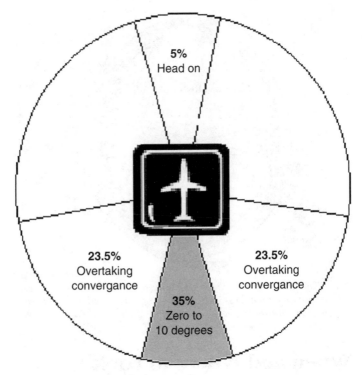

7-2 *Convergence angle diagram.*

These statistics provide important clues for a visual scan pattern. From a defensive flying standpoint, we are limited in what we can do by our field of vision, because it is very difficult to see and avoid someone overtaking us from behind. Emphasis on timely and accurate radio calls, especially in and around the airport, becomes all the more critical.

From purely the see and avoid standpoint, we must be more aware of the possibility that we will be the overtaking aircraft, and develop our scan pattern to mitigate this risk. There are basic scan patterns that will help us accomplish this task.

Scan patterns

There are no perfect scan patterns, but in general the concept is to move your eyes systematically to search an area roughly 60 degrees either side of your flight vector and 10 degrees above and below your projected flight path on a regular basis. Most experienced pilots recommend a "block" or "sector" scan as a foundational part of their composite cross-check. This is accomplished by pausing briefly to view each area of your scan, and to focus on a distant object. This avoids what is known as "fixed-field myopia"—the tendency for the eye to relax its focus to just a few feet in front of you in the absence of an object to focus on. Occasionally, you need to also make a conscious effort to "look around" any fixed obstructions, such as wing struts or opaque portions of the cockpit, and to clear areas outside the main scan area.

Once an aircraft or hazard is sighted, a pilot must quickly determine if the aircraft is on a collision course in order to determine the need for evasive action. Usually this can be accomplished by looking for relative motion. An aircraft that is on a converging course and is not moving in your field of view is likely going to be a conflict. That is not to say that aircraft that are not currently a conflict can be discarded, because any aircraft in your vicinity can quickly become a conflict if either aircraft alters its flight path. Don't assume you can predict what the other guy is going to do, even if he has made a radio call stating his intentions. In the words of Ronald Reagan, "trust but verify."

Even if we do everything right with our scan pattern, there are still a few pitfalls to be aware of when operating under see and avoid conditions. Among the worst is the degradation of visual acuity because of atmospheric conditions such as fog, haze, and glare.

Case Study: Glare and a Midair (NTSB Report Number ANC97FA051A)

"On April 9, 1997, about 1255 Alaska daylight time, an amphibian float equipped Cessna 185 airplane, N59AK, and a float equipped Bellanca 8GCBC airplane, N5025K, collided in midair about 12 miles northwest of Tatitlek, Alaska. The Cessna received minor damage to the right amphibian float; the commercial pilot and the private pilot passenger (observer) aboard were not injured. The airplane continued on to Birchwood Airport, Birchwood, Alaska, and landed without incident.

"The Bellanca airplane was destroyed, and the two occupants, a commercial certificated pilot and a passenger (observer), were fatally injured. The Bellanca fell into the waters of Prince William Sound, and the victims were removed by nearby fishing boats. The two airplanes were operated under 14 CFR Part 91, and visual meteorological conditions prevailed. No flight plans were in effect for either airplane.

"The pilot of the Cessna, and the observer, were interviewed on April 9th by the NTSB investigator-in-charge (IIC) at the Birchwood Airport. The pilot reported he had started flying about 0900, and had stopped for lunch and fuel at a fishing vessel in Galena Bay. After the stop, he resumed his fish spotting duties. The pilot stated he and the pilot of the Bellanca airplane were among *a group of 20 or more airplanes all flying in the same general area* [emphasis added] near Galena Bay to spot herring (the airplanes fly in support of nearby fishing vessels, and the pilots, or their observers, radio information to the boat crews where the herring are schooling. The herring fishing season had not opened, but an opening was deemed imminent). The other airplanes *were dispersed in about a one mile circle, with altitudes estimated from 1000 feet msl to 3500 feet msl* [emphasis added].

"The Cessna pilot related his airplane was at 2000 feet msl, in level flight, with 20 degrees of flap extended. He said his airplane's wingtip and vertical stabilizer strobes were turned on, as well as the navigation lights. His airplane was heading in a southwesterly direction, exiting the mouth of Galena Bay. The pilot said he was flying the airplane from the left seat, and that his passenger, whom he employed as an observer, and who is also a private pilot, was seated in the right front seat. He said neither he or [sic] the observer saw the Bellanca until an instant before impact, when they saw 'a flash of yellow.' He said he believes the collision was nearly head on, with the Bellanca slightly below and to his right. He said the impact happened so quickly that he does not recall a conscious attempt to avoid the Bellanca, but that he probably 'instinctively' pulled back on the control yoke.

"After impact, the pilot said he became aware that the other airplane had crashed. He radioed for a nearby airplane to fly alongside of him and assess the damage to his airplane. The damage appeared to be limited to the right amphibian float, and a decision was made to fly the airplane to Birchwood (approximately 110 miles west of the accident site) for landing and subsequent repairs.

"The observer in the Cessna, a private pilot, related essentially the same information as the pilot. The observer said his job was to look for fish and other aircraft, assist in the communications, and to take over the controls if required in an emergency. He said he had no flying duties, and that he did not see the Bellanca until an instant before impact.

"The pilot of the Cessna was a member of the Alaska Fish Spotter's Association, and has been spotting fish for several years. *He was involved in another midair collision in 1991 near the same location (Tatitlek). That accident also occurred on April 9. There was a single fatality in the other airplane.* The pilot of the Bellanca was also a resident of Homer, was a member of the Alaska Fish Spotter's Association, and had several years experience spotting fish in Alaska.

"*Meteorological information:* The pilot of the Cessna indicated in his verbal statement to the NTSB investigator-in-charge that at the time of the collision, he was flying the airplane in a southwest direction. In his written statement to the NTSB, he related he was flying in a westerly direction, about to exit Galena Bay. He also noted in his written statement that his visibility was hampered by bright sun reflecting off the ocean and snow covered terrain.

"At the time of the accident, 1255 (estimated) Alaska daylight time, a review of solar tables places the sun at an azimuth of 163.6 true (138.6 magnetic), and 36 degrees above the horizon. Witnesses, and the pilot of the Cessna, described the prevailing weather as clear and sunny, with light winds.

"The Cessna pilot stated that at the time of the accident, he was monitoring the State of Alaska's Fish and Game radio frequency, as well as the common traffic advisory frequency the other fish spotting airplanes were using. He said he was not in communication with the pilot of the Bellanca. It is unknown what frequency(s) the Bellanca pilot was monitoring. Both airplanes were operating in Class E, General Controlled Airspace. They were not in contact with any air traffic control facility, nor were they required to be. At the altitude they were flying, radar coverage was not available in the accident area."

The National Transportation Safety Board determines the probable cause(s) of this accident as follows: "The command pilots of both airplanes inadequate visual scan, which led to their failure to see and avoid each other's airplane. A factor associated with the accident was sunglare."

Preparation steps that might have avoided this mishap

This tragic mishap provides a clear example of an extremely high-risk operation where previous mishaps had occurred and lessons were available. The fact that the pilot of the Cessna had himself been involved in another midair fatal mishap on the same date in the same area, shouts for corrective actions to deconflict these aircraft. Clearly, the combination of 20 aircraft circling in a 1-mi area compressed into a roughly 2000-ft altitude block all looking down at the water for fish strains common sense—when coupled with the reflective surface of calm seas and a 36-degree azimuth sun glare—it exceeds it.

In-flight actions

It appears from the passenger briefing that the observer had partial responsibility for clearing for traffic, but according to both pilot and observer, they didn't see the Bellanca until it was far too late to maneuver. To be fair, the westward flight path of the Cessna into the sun places much of the see and avoid capability on the Bellanca crew, who would have had less glare to deal with, but it seems as if the mission may have gotten priority over safety in this case from the crews of both mishap aircraft. Short of leaving the area or climbing away from the traffic until the aircraft vector was more conducive to see and avoid (away from the glare), there was little to be done once the risk of the mission was accepted.

Lessons learned

One of the clear lessons in this case study is the need for a better system for the deconfliction of aircraft in this extreme setting. Beyond that, the realization of the glare phenomenon should make everyone more vigilant, but especially during periods when you might be a conflict from an aircraft who's see and avoid capability is seriously compromised, that is, when flying toward an aircraft that is flying into the sun.

"Seeing" with Your Ears

Every pilot knows that often it is your ears that direct your eyes. Paying close attention to radio calls from air traffic controllers and other pilots helps provide the three-dimensional picture of what is going on around you—and what might likely occur in the future. By developing the skills to maintain good situational awareness (SA), pilots can improve their odds when operating around other aircraft. This can be especially helpful in a highly congested airport traffic area, a lesson these two instructors and students are unlikely to forget.

Case Study: Congested Pattern (NTSB Report Number NYC91LA015B)

"The two Cessna 152's collided while operating in a congested airport traffic pattern. There were 7 airplanes in this pattern at the same time. The airport has no control tower. The Kent State University flight training coordinator said, 'there was a great deal of transient traffic this day...the radio frequency was saturated with radio calls.' The pilots lost visual contact with each other, because there were so many airplanes and some of these were not familiar with the normal traffic pattern, after the midair collision, both airplanes were able to land safely."

The National Transportation Safety Board determines the probable cause(s) of this accident as follows: "Failure of both pilots to maintain visual separation in the landing traffic pattern. Factors that led to this midair collision were the excessive number of airplanes in the airport traffic pattern and radio frequency congestion."

Clearly the two instructors are ultimately responsible for the safety of their aircraft in this environment, but their comments to the investigators illustrate two key points:

1. Airfields that are utilized primarily for flight training are particularly susceptible to midair collisions. There are several possible reasons for this, including the split attention of the instructor that leads to reduced SA, and the frequency of "nonstandard" pattern ops from students who are still learning the ropes.

2. Even at the "home airfield" you do not have exclusive rights, and those who are transients might not follow "standard" operating procedures for a flight-training faculty.

Once again this case study highlights the need for predictable communications and constant vigilance in the traffic pattern. The next case study takes this one step further, and points out that even in so-called controlled airspace a pilot's best defense is the eyes and ears.

Case Study: When Things Begin to Unravel (NTSB Report Number CHI93IA167)

In the following incident, two pilots were put in a precarious position, but were able to avoid a midair collision in spite of the controller's mistakes. Let's see what we can learn from their actions.

"*Synopsis:* The local controller cleared a Piper PA-28 for touch-and-go on Runway 15, and shortly thereafter, cleared a Cessna 500 corporate jet for takeoff on Runway 10. The two Runways were situated so that the extended centerline of Runway 10 crossed the end of Runway 15. Witnesses indicated that the two airplanes passed within 100 to 300 feet of each other. The pilot of the Cessna reported he visually acquired the piper as both airplanes were in their respective takeoff rolls. The local controller was communicating with three airplanes at the time of the incident. He reported he became distracted by inbound traffic and neglected to adequately monitor the departing traffic. He said he should have issued a full stop landing clearance to the Piper instead of the touch-and-go clearance.

"*History of Flight:* On May 10, 1993, at 1011 central daylight time, a Cessna Citation 550, N41SM, had a near midair collision with a Piper PA-28, N9291W, while operating in the Du Page County Airport Traffic Area, West Chicago, Illinois. Neither of the two airplanes sustained damage. The two airline transport pilots and three passengers aboard N41SM were not injured, nor were the two pilots of N9291W. Both airplanes operated in visual meteorological conditions under 14 CFR Part 91.

"An FAA Air Traffic Control Tower was operating at Du Page Airport at the time of the incident, and both pilots were communicating with the local controller. At

1009, N9291W was cleared for a touch and go landing on Runway 15. At 1010, N41SM was cleared for takeoff on Runway 10. At 1011, N41SM passed over N9291W just as N9291W was lifting off. The two airplanes crossed near the extended centerline of Runway 10 and the centerline of Runway 15. Vertical separation was estimated as 50 to 300 feet; lateral separation was estimated as 0 to 100 feet.

"The pilots of N41SM said they were given takeoff clearance from the tower controller and subsequently obtained visual contact with the touch and go Piper on Runway 15 when they were approximately ½ to ⅔ into their takeoff roll. They were unaware of the Piper's presence until that time. They determined that the Piper might present a collision hazard, and were able to keep the Piper in sight during the remainder of the takeoff and initial climb. Their joint crew statement reads, in part: 'We maintained visual contact with the traffic on Runway 15 as the pilot rotated. As we rotated and climbed at V2 and Runway heading, we were able to maintain visual contact with the other aircraft until well clear.'

"The pilot of the Piper airplane (N9291W) is a certificated flight instructor who was giving a biennial flight review to a private pilot. The instructor's statement reads, in part: 'Turning left downwind to base on what I believe to be our fourth touch and go when we received clearance from tower controller for touch and go...I heard the controller give caution to a Citation departing 10 (Runway). I looked out my side window and there was a Citation headed towards us as we lifted off.... The Citation made a correction towards the northeast momentarily then over us, which put him maybe 300' above and behind us.'

"The Cessna Citation (N41SM) was departing Runway 10/28. Runway 10/28 is 4,751 feet long and 75 feet wide.

Piper N9291W was touch and go on Runway 15/33. Runway 15/33 is 3401 feet long and 100 feet wide.

"At the time of the incident, there were no recorded anomalies with any of the air traffic control facilities, or any current notices to airmen pertinent to the operation of either incident airplanes on their respective runways. The transcript indicates the controller was actively working three airplanes: the two incident airplanes, and a Cessna 152, N389GT, inbound from the west. N389GT originally reported 'about ten east,' which was later clarified to be west. N389GT was cleared to land on Runway 10 at 1008. N9291W was cleared for a touch and go on Runway 15 at 1009. The Cessna Citation, N41SM, was cleared for takeoff on Runway 10 at 1010. The local controller issued the following instructions to N41SM at 1011: 'Cherokee niner one.... Citation one Sierra Mike start a right turn immediately Cherokee traffic ahead.' The pilot of N41SM responded: 'Roger we're going to have to go over him here.'

"Immediately following the incident, the tower chief relieved the incident controller from duty. The incident controller subsequently was granted administrative leave, and was decertified from the full performance level pending retraining. The incident controller was interviewed by the NTSB investigator on May 11. During the interview the controller was asked how the loss of separation between the two airplanes occurred. He responded with words to the effect that he had become distracted with the inbound traffic, and that he should have issued instructions for the touch and go Cherokee to make a full stop on Runway 15."

The National Transportation Safety Board determines the probable cause(s) of this accident as follows: "Inadequate control tower service by the local air traffic controller by providing conflicting clearances to the pilots of two aircraft."

Pilot actions

It should be clear to the reader that the pilots on both of these aircraft were in complete control of this situation. It should also be clear that they needed to be to avoid a mishap. From the limited testimony that we have available, it seems as if both aircraft were aware of the conflict early, and both had visually acquired the other. This is likely because of solid preparation for this particular field and some predeterminations on how to handle events during takeoff roll.

As this event unfolded, each of these flight crews took positive actions without panic or causing additional hazards. Luckily, the Citation had the performance capability to climb over the conflict, and in so doing avoided a midair collision.

Lessons learned

The most important lesson to be learned from this event is that any time you are in visual conditions, do not rely solely on the controller for your separation. A second lesson can be gleaned from the professional manner in which the Citation pilots made their evasive action. By climbing straight ahead, they did not confuse an already deteriorating situation, and relied on their own air sense to resolve a dangerous airspace conflict.

Cockpit Resource Management (CRM) and Midair Avoidance

There is an old saying that "two heads are better than one." The midair collision avoidance corollary to this might be "Four eyes are better than two—and six or eight are even better than that!" Pilots are often hesitant to brief their passengers on what safety assistance they can provide. This is a big mistake for two reasons. First, many, perhaps even most, passengers like to be involved in the flight operation. It gives them a feeling of contributing to the flight. Second, some pilots don't want passengers interfering with flight operations. A solid CRM briefing can take care of that.

CRM briefings should include the potential for conflicts with other aircraft, where and when they are most likely to occur, and how to call out traffic to the pilot. Additionally, some mention of the dangers of time compression and distraction should be included and perhaps quick overviews of what paper resources are available in the cockpit, such as charts, approach plates, and the like. A good CRM briefing is like an insurance policy, you hope that you won't need it, but are damn glad to have it if required. In the following example, CRM broke down, then reemerged in the nick of time to save a deteriorating situation.

Case Study: Better Late (CRM) Than Never

"The cause of the near midair was attributed primarily to a breech of normal ATC procedures in that the controller failed to provide appropriate traffic advisories to West Wind (N767AC) for recently departed traffic (Archer N9112Z) on Runway 30 at Palwaukee. A secondary cause was ATC (tower controller) creating an untimely diversion to N767AC by advising the pilot during takeoff of a situation of a non-emergency nature having to do with his aircraft not receiving fuel that he paid for. This action caused the pilots to breech a sterile cockpit environment and break normal cockpit events. This cockpit diversion caused the flying pilot to divert his attention inside the cockpit along with the non-flying pilot. The non-flying pilot discovered the imminent collision and took the flight controls, avoiding the other aircraft. The pilot in command of N767AC was verbally counseled on CRM and proper collision avoidance. The pilot of N9112Z did not see or know of the near midair."

Once again the big sky theory prevailed (big sky + little aircraft = safety), but not without a little help. Although the crew was initially distracted, something prompted the pilot not flying to look outside, and he felt the need— and authority—to momentarily take the aircraft from the captain to avoid the collision.

Collision Avoidance Checklist

The AOPA has developed a nine-point checklist that it recommends as a guideline for avoiding midair collisions.

- *Plan your flight.*   Know your route, the frequencies you'll need along the way, and the airport you'll be arriving at. Fold charts and preset navigational aids to maximize scan time. Program GPS and Nav computers on the ground to minimize heads down time in the air. Consider where high traffic/high workload areas will be. Avoid them if possible or plan on being extra vigilant during those phases of the flight.

- *Clean your windshield.*   A squashed bug on the glass can block an aircraft from view and make it more difficult to focus properly. Make S turns for improved forward visibility during the climb. Climbing at cruise airspeeds accords a better view over the nose.

- *Enlist passengers.*   As part of your preflight briefing, explain basic scanning procedures to passengers and have them assist in looking for traffic. Explain FAA radar advisory procedures, so they can help locate traffic called by ATC.

- *Use aircraft lights.*   Install and use anticollision lighting so that if you don't see a potential threat, maybe the threat will see you. Strobe lights can improve an aircraft's ability to be seen day or night. Use your landing light on approach, departure, and climbout.

- *Use sunglasses.* Sunglasses that block out UV rays help protect your vision and reduce eye fatigue. Red/yellow spectrum lenses make it easier to see through haze. Polarized lenses reduce glare, but this may be a detriment to spotting traffic as the glint of light bouncing off an aircraft may help make it visible.

- *Observe proper procedures.* Use correct cruising altitudes and traffic patterns. Announce your position at nontowered airports. Recognize that not everyone follows the rules.

- *Communicate.* At airports with radar approach control, contact the facilities at the distances prescribed on aeronautical charts. At nontowered airports, announce your position starting 10 mi out.

- *Become a target.* If you operate an aircraft without radios or transponders, consider installing them. Regulations require that aircraft equipped with transponders must squawk mode C while in flight.

- *Scan for traffic!* Use the techniques presented. Adapt them to your needs. Use your eyes and your head together.

A Final Word about Technology

Many will take note that this chapter spoke very little about collision avoidance technologies. This was by design and is not intended to denigrate the outstanding systems that have been developed. In many cases, systems such as TCAS and ADS-B are lifesavers. But technology changes, and machines fail. Overreliance on any system that is created by the imperfect hand of man needs a human backup. The entire Controlling Pilot Error series is oriented toward human solutions.

Summary

As we have seen throughout the rest of this book, there are many errors that can occur in the approach and landing environment, but most play out their end game in the landing phase. Midair collisions are different— they can happen anywhere at any time. There are a few key points we have made in this chapter that I want to reinforce with a brief summary.

1. *Keep your head outside of the cockpit.* In today's high-tech cockpit, there are plenty of toys to distract us. There is no substitute for a disciplined visual scan.

2. *Develop a disciplined scanning technique.* Know where you are looking and why. Develop your own technique or talk to other pilots about theirs, just don't ignore the need for a well-thought-out scan pattern.

3. *Practice good CRM.* Use additional crew or passengers to assist in clearing and to help avoid distractions. Keep a sterile cockpit to avoid distractions when in the high midair risk areas.

4. *Keep your sensors sharp.* No matter how good your technique is, if your eyes are failing you will miss critical visual cues. If you need glasses, wear them—all the time. Stay rested and fit, and your vision should perform as your most reliable safety device.

5. *Stay vigilant and leave yourself an "out."* No matter how safe you feel, the potential for a

conflict is always there. Keep yourself ready to handle the unexpected and have a plan when it does occur.

The traffic pattern is the most likely location for a midair collision to occur. As we have seen in this chapter, there are several keys to effective separation in the airport environment. While all are important, perhaps it can be summarized in one statement—*be where you are expected to be*. This leads us to our next chapter where we look into the missed approach, one of the most overlooked skills in a pilot's approach and landing repertoire.

8

Missed Approach: Routine Action or Your Last Chance*

*This chapter was written with Noel Fulton.

You've flown the perfect approach. The weather is reported just above minimums and you are 100 ft above decision height (DH) on speed, on course, on glide slope. You've got this baby wired and you're just waiting for the runway lights to start blinking through the undercast. But something is not right—you should see something by now. As the altimeter slides through the DH, you realize that you are not going to break out and you transition to the missed approach. Are you ready? If you are, this is just an extension of what you've been doing all the way down final approach—a controlled and stabilized maneuver. If not, it may be the most difficult maneuver of your flight in the worst possible environment—the high-density traffic area. If it comes at the end of a poor approach, it may be your last chance to prevent an accident.

If you forget everything else in this book—if you forget about approach planning, stable approaches, or aircraft performance—you can still go missed approach. It is a get out of jail free card. It's "live to fight another day." It can save your life. It's your last chance to turn a possible accident report into merely the "worst approach I ever flew." Let's look a bit closer.

A Change of Attitude and Language

We expect to land from nearly all of our visual approaches and the vast majority of instrument approaches. After all, we're all good pilots, right? We certainly plan our approaches methodically. We get nice and stable. We communicate effectively to ATC and our flight crew and take performance issues into account. So naturally, we expect to put the aircraft down on the runway. But maybe we shouldn't. Maybe our mind-set is setting us up for a poor execution of a missed approach if it becomes required.

I remember during instrument training, my instructor told me to plan to go missed approach and be surprised if I landed. When training under the instrument hood that was certainly true. You never knew if your instructor was going to have you "break out" or not. You were ready for the missed approach because you didn't know what was in your instructor's game plan for the day. But unless I'm on a training ride or check ride, I still expect to land.

Listen to our instrument approach briefings. In general we brief all of the required information: the frequencies, the headings and altitudes, and the runway environment. Finally, at the end, when the approach to expected landing has been covered thoroughly, we say something to this effect: "In case we go missed approach...." We say "in case" because we expect to land.

No matter the advice of our instructors, check airman, or even common sense, the missed approach is the surprise, not the landing, and the language in our briefings subtly and powerfully reinforces this every time we brief. Because we expect to land and our briefs strengthen this expectation, it takes something drastic to move many—if not most—pilots off their expectations. How many times have you heard of pilots going below a minimum altitude trying to break out of the weather with disastrous results? We've heard this far too many times and we usually shake our heads. But we also understand the surprise at not breaking out and the temptation to keep going "just a little further" because we've all been there at the decision height a little shocked when the runway doesn't appear.

Pilots need to change their expectations. We need to begin expecting those missed approaches when there isn't an IP or check airman on board. The beginning of changing this deeply entrenched expectation is to change how we brief our approaches. We should switch the touchdown and the missed approach sections of the brief. Instead of briefing the touchdown and landing environment right after we brief the minimum descent altitude, decision height, and missed approach point, we should brief the missed approach. For example, we should say something like, "Once we reach the decision height of 632 ft MSL, 200 ft AGL, we will *plan on* going missed approach by climbing on runway heading to 1500 ft MSL then turning right direct to the VOR to enter holding. *If we break out and things look safe to land,* we will expect to see Runway 35, which is 10,000 ft long and 150 ft wide." Do you see the difference in switching the brief and the expectations around? Briefing an approach in this way reinforces the more conservative option of going missed approach. It puts expectations on the side of taking the aircraft around instead of trying to salvage a bad approach or pressing below minimums. And if we have done everything right, and things look safe to land, we are ready for that too. We have added the conditions for landing that will help us make a good decision.

Salvage Operations

Most pilots have salvaged a bad approach. We may start off a bit high or a bit fast, but eventually, we are able to get stable and land through skill, cunning, or just plain dumb luck. A large part of pilot training in the pattern or during instrument training focuses on fixing the deviations all the way through the approach and the rollout. It has become second nature. Unfortunately, our skill at salvaging minor deviations doesn't necessarily translate to salvaging seriously screwed up situations.

Often, the best way to salvage a bad situation is to take the aircraft around instead of trying to put it down. Unfortunately, pilot culture commonly views this as "not being able to hack it." Never mind that you should have been stable and configured long before things officially went to hell, "good pilots" can salvage bad approaches— or so say many. By trying to salvage a poor approach, you are putting yourself in a situation where you have to do everything right to save the day. And although you may succeed 99 times out of 100, Murphy is patient and will wait for that one time when you are fatigued or complacent and he will get you. The NTSB will write your epitaph, usually something like this: "PIC descended below MDA without adequate visual references."

Isn't it better to salvage the bad approach by taking the aircraft around before you have to do "everything" right to get it on the ground? Pilots understand the concept of "safety margins." The aircraft operating limits allow a margin of safety for the equipment. The regulations do the same for pilots. Instrument approach procedures and missed approach instructions ensure a margin of air between you and the hard stuff on the ground. If all of these aspects of aviation have wisely built in safety margins, why would you go right up to—or past—the limit of your skills when confronted with a difficult landing situation? I'd like to go around before perfection was necessary for my survival because I am all too aware of the small mistakes I make. Go ahead and salvage the minor deviations as you have been trained, but when things get bad close to the ground, go around and try again.

Case Study: Missing the Runway (NTSB Report Number LAX97FA036)

"On November 14, 1996, a Cessna 310I collided with five unoccupied parked airplanes during initiation of a missed approach at the Van Nuys Airport, Van Nuys, California. The airplane was destroyed, and the airline transport pilot was fatally injured. Instrument meteorological conditions prevailed, and an instrument flight rules (IFR) flight plan was filed. The on-demand air taxi flight was operating under 14 CFR 135. The flight originated from North Las Vegas, Nevada.

"The flight started fine under VFR conditions and was intended to deliver some bank cargo. While cruising on the beautiful day over the desert and mountains of Southern California, the pilot received word that the weather at his destination was deteriorating and was, in fact, below landing minimums. The pilot decided to divert to the Van Nuys airport. While enroute to Van

Nuys, the weather there went IFR and the controller issued the mishap pilot an IFR clearance and permission for the ILS approach to runway 16R. The pilot acknowledged the clearance and commenced the approach.

"As the pilot approached the airport, ATC relayed two recent PIREPS stating that the weather at Van Nuys was actually VFR. The pilot acknowledged again and continued the ILS approach. The pilot switched to the Van Nuys advisory frequency when directed and was asked to report canceling IFR.

"The pilot transmitted that he had the airport in sight but did not cancel his IFR clearance. Radar showed his aircraft at 2000 feet at the time of his transmission. According to radar information, the airplane continued to descend as it approached the airport and reached the middle marker at decision height. Radar further indicates that the pilot crossed the north end of the airport about 100 feet AGL and began a climb thereafter.

"Witnesses on the ground reported seeing an aircraft lined up on the runway disappear into ground fog at about 75 feet above the ground. No witnesses reported seeing the aircraft hit the ground. The wreckage of the aircraft was found on the tarmac and the cause of the accident appeared to be the mishap pilot's failure to maintain a climb during the missed approach due to spatial disorientation caused by the fog. Further investigation revealed a previous mishap by this pilot and a termination for poor instrument flying skills within the previous year."

Analysis

This is a textbook case of a pilot who wasn't prepared for the missed approach. It was a diversionary field, and it may be unlikely that the mishap pilot spent much time studying the approach plate when the weather for the flight started off in VFR. It's also likely that the nature of this foggy evening lured the pilot into a more dangerous situation. The pilot reported seeing the runway at 2000 ft and a few miles. He naturally started down thinking everything was fine. Low-altitude ground fog can be very surprising. Even though the pilot was surprised, a properly executed missed approach would have easily avoided the mishap and saved this pilot's life.

The mishap report concluded that spatial disorientation (SD) caused the pilot to descend and hit the ground. But what set the pilot up for the SD episode? Is it possible that the pilot had not reviewed the missed approach procedure, and was desperately searching out the window for another visual clue? Perhaps he caught sight of the airport as he flew over the field and tried again to descend. This pilot did not have the strongest skills, as indicated by his firing a few months previous to this accident along with a previous mishap. A missed approach would have helped this pilot live to improve his skills. The key to a successful missed approach begins with knowing when to execute it.

The Missed Approach Point (MAP)

It is so basic it shouldn't have to be repeated. When they design a missed approach, they design it to be flown from the missed approach point. If you delay your missed approach procedure for any reason, you risk everything. We don't have to beat a dead horse here. If you do delay, maybe you'll get lucky like the pilots in the following two case studies. But if Murphy is paying attention, you'll end up like the pilot in the third study.

Case Studies: Lucky, But Not Good

"While flying a GPS approach I failed to initiate a missed approach when I realized that I had missed the airport. I was underneath the clouds hunting for the field when I finally returned to the controller and declared a missed approach. I was given a 090 heading but turned to 270 instead. Then I turned 090, then flew 180 and came around for another approach that went fine. I contribute this incident to not flying the missed procedure when I realized I was lost, inexperience with GPS approaches, poor judgment and getting behind the approach." (ASRS Accession Number 420894)

"In June of 1998 I had scheduled an instrument training flight with my flight instructor to fulfill my long cross-country requirement for an instrument rating. I was in the left seat and my CFI was in the right seat. Since the weather was VMC, I wore a hood and flew the airplane for everything. The departure and the enroute portions of the flight were uneventful.

"I received radar vectors for the ILS 9 approach to Duluth where conditions were 200 feet overcast and visibility one mile at night. We were above the clouds while being vectored and heard another aircraft attempt the ILS approach. I got established on the approach and was still in VMC and headed down the chute. The checklists were complete and I felt familiar with the approach and the missed approach procedure. I began to struggle with the approach overcorrecting on both the course and the glideslope. At approximately two miles out, I hit full-scale deflection on the course needle and went missed. I began climbing, reported missed to ATC and, at my IP's request, asked for another ILS approach.

"I was given a trip around the box and again found myself on course at the outer marker. I made smaller corrections this time never getting more than a dot on the localizer and staying pretty much on the glideslope the entire time. I eventually did get about a dot and a half deflection on the course needle and couldn't seem to get back on course. I went missed again. I thought I was climbing out straight ahead as the CFI pointed out the flashing approach lights well to our right. I saw the lights but continued to climb straight ahead. A few seconds later (I think) I heard a loud 'thunk' and saw something flash to my left. I immediately pulled back and climbed out in a slight left bank. I leveled the wings and continued to climb to 2000 feet MSL. Tower radioed that we were not following the published missed approach procedure. I asked tower to stand by and they told us to keep climbing. I then declared an emergency saying that I thought we had hit a tree.

"We continued to climb and assess the damage. I made numerous turns using the controls and asked ATC for vectors to the nearest VFR airport, which turned out to be about forty-five miles away. I calmed down as we went since we didn't think there was any significant damage to the plane. We thoroughly inspected everything we could with our flashlights. The only noticeable damage seemed to be the left strobe and nav lights. They wouldn't come on. The flight to the VFR field was uneventful and we had time to discuss the incident. We think that when the IP pointed out the airport to my right, I continued to descend into the terrain even after I had commenced the missed approach. We really lucked out that the damage was so minor." (ASRS Accession Number 405980)

Analysis

These pilots used up three lifetimes' worth of luck. They broke a cardinal rule of instrument flying and luckily lived to tell the tale. The first pilot admitted he was searching underneath the clouds and completely messed up the missed approach procedure. He certainly wasn't prepared for the missed approach, which may have been why he delayed it. He was unfamiliar with his equipment and the procedures of GPS approaches. These weaknesses in his overall airmanship could have easily killed him. Consciously addressing those weaknesses during thorough preparation may have avoided what luckily only turned out to be an ATC incident.

The second two pilots seemed to actually hit a tree in their delayed adventure. They allowed themselves to get distracted during a critical phase of flight. Do not allow yourself to be distracted when you are low, slow, and going around. Treat every missed approach seriously and fly the procedure as precisely as you can. Maybe there isn't anything nearby that could kill you, but it is not the pilot's place to question the missed approach procedure in the air. Fly it! If you have questions, save them for when you are on the ground. Don't rely on luck to save you if you forget basic instrument rules because luck can desert you in a flash.

Case Study: Deadly Delay (NTSB Report Number NYC96FA163)

"It was a standard IFR flight from Charlotte, North Carolina to Roanoke Regional Airport in Virginia. The weather at Roanoke was foggy with rain on a dark night. The mishap aircraft was a Beech A36 on a cargo flight with a commercial rated pilot at the controls. The Beech was destroyed and the pilot killed when it impacted mountainous terrain during an instrument approach. This was controlled flight into terrain. The pilot may have been nervous at the last second but he probably died relaxed.

"Air Traffic Control tapes show that the pilot was cleared for the ILS 33 approach to Roanoke which would have the aircraft flying northwest as it approached the airport. The airport was closed and the mishap pilot was communicating with another aircraft coming in behind him on the common traffic advisory frequency.

"Radar data shows that the Beech was about 6 to 7 miles southeast of the NDB at 6200 feet. The controller terminated radar services and the Beech crossed the NDB at about 5800 feet continuing down to the outer marker at 4900 feet, the middle marker at 4700 feet and the airport at 4100 feet. Radar showed aircraft continuing northwest from the airport while still descending. The last radar contact showed the Beech on a five mile bearing from the Roanoke field at an altitude of 2700 feet.

"During the approach, the commuter crew coming in behind the mishap aircraft asked the accident pilot over the Roanoke CTAF what altitude he broke out of the clouds, which the pilot responded, '...were still about twenty nine hundred, we're starting to see some ground now.' About 25 seconds later, the accident pilot radioed, 'I'll tell you what I'm gonna do is I'm gonna do a missed approach, my uh needles are (?) working (?).' This was the last recorded transmission, which ended abruptly as the aircraft slammed into the hillside. Two witnesses on the ground reported hearing an aircraft pass overhead with the engine sounding like it was at low power. One witness even heard the impact.

"The published instrument approach procedure for the ILS runway 33 approach to Roanoke depicted the airport elevation to be 1176 feet, with 'mountainous terrain higher than airport in all quadrants.' Runway 33 was 5800 feet long and 150 feet wide. The profile view depicted a glide slope intercept and minimum altitude of 3800 feet prior to the Vinton NDB, followed by a descent to the decision height of 1660 feet. The Localizer runway 33 approach profile view depicted a descent from 3800 feet after crossing the Vinton NDB, with a minimum altitude of 2600 feet until crossing the Outer Marker Beacon, followed by a final descent to the minimum descent altitude of 1940 feet. The missed approach point was based on a groundspeed and elapsed time from the Vinton NDB. The missed approach point (MAP) for the ILS runway 33 approach, was located prior to reaching the approach end of runway 33. The MAP procedure required a climbing left turn to 4000 feet, and depicted a holding pattern at an intersection, southeast of Roanoke. Again, the aircraft hit terrain five miles northeast of the field."

Analysis

This is pretty clear-cut. The pilot seemed to lose situational awareness on the approach. He flew past the missed approach point while still descending into mountainous terrain. Going back to planning the approach, review of the airmanship model would have highlighted the high terrain in all quadrants along with the poor weather under the environment pillar. These factors would have also been important when considering the risks in this approach. I'm not saying that the risks were so great that this pilot shouldn't have attempted this approach. But the review may have given the pilot enough skepticism to be primed to go missed in that environment. This pilot, like all pilots, expected to break out and land. He kept flying northwest well past the missed approach point searching for the field. He just didn't have the luck of the pilot above. Maybe with a thorough consideration of the risks, and a plan that included an expected missed approach rather than an expected landing, he may have lived to fly again.

The following case studies don't show perfect aviators because there is no such thing. They show human beings doing their best and, ultimately, making good decisions.

Case Study: Could Have Been Much Worse (ASRS Accession Number 427590; Narrative by Captain)

"It was the last leg of a very long thirteen-hour day. We were flying a charter flight into Syracuse. It was very late and we were expecting to shoot the ILS approach to runway 28. I planned the descent poorly and had to really dive down to get to the initial altitude upon being cleared for the approach. We finally captured the localizer but were way above the glideslope and continued to rapidly descend to catch it. We were at least configured and all of the checklists were complete, however, as we approached the outer marker, it became apparent that we would not be established on the glideslope. We went ahead and executed the missed approach as published climbing to 3000 feet. The tower told us to maintain 2000, as we were about to level off at 3000. We descended immediately and flew a heading the tower gave us. The next approach was uneventful. I contribute this incident to my somewhat low time in a new aircraft and an initially poorly planned descent. Fatigue was also an issue. Finally, I learned that proper planning and proper procedures are never more important than on the last leg of a long day."

Case Study: A Good FO (ASRS Accession Number 381063; Narrative by Captain)

"I was the captain on an aircraft flying from Chicago to Piedmont Triad International. We were being vectored for the ILS runway 23 approach and on a downwind vector about abeam the field when we were given a 200 heading to intercept the localizer. During the turn, it became obvious that we were too close and going to shoot through the localizer, so the controller had us turn further right to a heading of 260 to intercept the inbound course of 232 which was approximately a 30 degree intercept from the other side.

"We were told to maintain 3000 feet until established and I increased the intercept to about 40 degrees to expedite the intercept and started down since I was high on the glideslope. The FO reminded me that I was not yet established and I leveled off until the course came alive. The course needle jumped off the side of the case as I blasted through it with my 40-degree intercept. The controller offered to give us an ASR approach or turn us back to the localizer to re-intercept for another attempt. I told the FO I thought we could salvage this approach and started down again while re-intercepting the course.

"At this time, I was two dots high on the glideslope so I dropped the gear, set the flaps and pulled the throttles back. I figured that I could drop through the forecast cloud layer at 2900 feet, acquire the field and land, since the weather was reported at 1900 feet overcast with visibility at 5 miles. I entered the overcast with the power back and about a 2000 foot per minute descent when I heard a Ground Proximity warning of 'TERRAIN!' followed by 'PULL UP!' I started to arrest the descent when the FO reminded me to execute the Ground Proximity

Warning procedure, which is basically an immediate full power climb-out. I pushed the power up and the FO further advanced the throttles to go around power. We were above 1200 feet AGL when we received the Ground warning, which was why I wasn't initially concerned.

"We notified the controller and he gave us an altitude of 3000 feet that I ended up overshooting due to the engines at max with the nose high for the go around. We were vectored around for another approach and I broke out at 400 feet above the touchdown and landed uneventfully. At the time of the incident, we had been on duty for twelve hours with seven and a half hours of flying with no adequate meal break during the entire day."

Analysis

The crews of these airplanes were certainly not perfect. They were forging links to an accident chain at an impressive clip. But the key is that before something really unfortunate happened, they broke that last link with a timely go-around.

The first crew flying the charter into Syracuse started down late. Good approach planning may have taken the crew fatigue into account and recognized the deteriorating situation a bit earlier. It would have saved a lot of trouble if the captain had communicated to the controller the need for some more space (or time) on the approach. The captain knew the approach was unstable and when the outer marker came up, they took it around. The last comment from this captain is worth repeating. "Finally, I learned that proper planning and proper procedures are never more important than on the last leg of a long day." That is an excellent lesson that we can learn the easy way from this case study.

The next scenario shows how easy it is for a captain to get fixated and how critical a good FO (or any observer) can be to the safety of our approaches. From the case study, it's apparent that there were some weaknesses in the individuals and the crew because of fatigue and possible hypoglycemia after flying the long day without an adequate meal break. Both of those factors are proven to slow cognitive processes and make pilots more tolerant of deviations. The captain says he knew the vectors from the controller were going to be tight but accepted them anyway. With a thorough review of the airmanship model and recognition of the fatigue, maybe this captain would have asked for more time and space on this approach.

The captain's situational awareness told him he was high but it was the FO who realized that they were not legal or safe to descend until the localizer course came alive. The captain was relying on a forecast that would have him break out of the clouds approximately 1900 ft above the touchdown zone, which is why he thought it was safe to descend aggressively once he received course guidance. Essentially, the captain put his faith in a forecast that was made by someone whose life wasn't on the line. Weather forecasters do their best and pilots appreciate the work they do, but that should not alter the way in which we fly our approaches.

In this case the weather was actually at 400 ft AGL and not 1000. Imagine if this captain had broken out at the 400 ft with the 2000-FPM descent! This may have been an NTSB accident file instead of a self-reported ASRS report.

The airplane's rapid descent set off the ground proximity warning. Most companies require their pilots to immediately go around if flying at night or in IMC and the warning activates. The warning and the recovery procedure are considered emergency actions. The following is the bold print/memory item procedure from

"In Flight Activation of GPWS Warning in IMC or at Night" in the Great Lakes Aviation Flight Standards Manual:

1. Wings: LEVEL.

2. Airspeed: 122 KIAS.

3. Power: MAX ALLOWABLE.

4. Gear and flaps: RETRACT.

5. Continue to climb at 122 KIAS until all visual and aural warnings cease.

While the captain started to "arrest the descent," the FO rightly called for the full ground warning go-around. When the captain just pushed the power up, it was the FO who selected maximum power to get the aircraft out of danger. This FO was consistently focused on the safety of the aircraft, and the captain seems to have been focused on the landing he expected. The aware and assertive FO seems to have prevented a possible incident or accident. Once again, after the missed approach is executed, the scenario gets anticlimatic. Both second attempts at landings above went "uneventfully" once the crews allowed time to get stable. A soundly executed missed approach is a routine action. A "salvaged" landing is not.

Case Study: Good Thing He Was There (ASRS Accession Number 365900; Narrative by FO)

"I was the FO on a flight from Denver to Kansas City. The captain was being checked by a Line Check Airman (LCA) for our company. I found out later that the reason for the check was because of `special tracking' due to a previous training problem. All pre-flight briefings were completed and the flight progressed quite normally until the beginning of the approach. The captain seemed very professional and gave detailed briefings, had solid knowledge and good command. The initial approach into Kansas City was well planned and well briefed. We ran into some icing and light chop on the descent to an ILS approach. Weather was 700 ft broken and visibility was 1½ mile with fog. Things looked standard, but everything changed once we were cleared for the approach.

"The captain was hand flying the approach and we ended up high on the glideslope with full-scale deflections (both left and right of the course). Airspeed was as high as 30 knots fast. At 1000 feet, the only thing stable was our configuration. Passing the outer marker, I stated that we needed to get stabilized now or abandon the approach. Since there was no response, I checked for incapacitation and watched the captain fly through another 100 ft unstabilized with no indication of a go around. I stated, 'Let's get out of here! Go around!' and within a few seconds the LCA in the jump seat also said, 'Go around.' I don't know whether the captain would have gone around on my command if the check airman hadn't backed me up.

"The missed approach was flown normally except for some altitude deviations at level off. I remember recommending that we turn the autopilot on so it could do some of the work. After again getting no response, I turned the autopilot on myself while announcing my actions. We were vectored around again and re-briefed the approach and accomplished checklists. We agreed to use the autopilot for a coupled approach. ATC gave us a heading of 080 but the captain turned to 180. I queried him and he said he thought ATC had said 180. The check airman said he heard 080. By the time we contacted ATC and found out that 180 was the correct heading, we were already flying it. The controller had missed my incorrect read-back of 080. The next approach wasn't great, but it was safe and we landed normally.

"I flew the return leg with the LCA in the left seat and the captain in the jump seat. In my opinion, I was primarily concerned with safety throughout the entire episode with my standing in the company a close second. I don't feel that I let the situation proceed beyond a safe limit despite a stressful approach and line check. But, as I mentioned earlier, what if the check airman wasn't there to reinforce my go-around call? I truly don't know when I would have physically interfered and I wondered if my actions were appropriate. The check airman said that they were although there were some things I could have done to help the captain out before things became too unstable."

It is somewhat embarrassing to be part of a crew when something like this happens. Situations like this do occur and many go unreported and others have worse results. Complacency can set in as a result of things we take for granted, such as good training, excellent equipment, and the competence of other crew members. In this particular instance, throw in an engine fire or some other problem and it could have gotten out of control quickly.

Analysis

This is an interesting case study. The narrator seemed to be a properly assertive and helpful first officer. The approach was properly briefed and all the checklists were completed. The problem appears to be the captain's weak skills and/or proficiency. Sometimes, preparation and stability aren't enough. Sometimes the pilot flying may even ignore your calling for a go-around because he is so fixated on trying to get the approach under control. So, what do you do then?

In this case, it's lucky that there was a check airman on board who added the weight of his experience and authority to the call of the first officer. The narrator wonders, as we do, what would have happened if the captain had not responded. Would you take the aircraft and initiate the missed approach?

The narrator mentioned that his concern for his "standing in the company" was a close second to his concern for safety. It's difficult to contemplate taking the aircraft from a captain when your job is on the line. The consequences seem so real and assured if we take that step. An accident seems far less likely; so unlikely, in fact, that the consequences don't seem as real as those for assuming control from the captain. I hope you are never put in a situation where you feel you have to choose between your life and your job, but there are other jobs. Make the choice now. If you are a first officer you may not have the knowledge and experience of your captain, but you are still responsible for the safety of your aircraft and any captain can have a bad day. That is why there are two of you.

Conclusions and Recommendations

Pilots expect to land from visual and instrument approaches. But, until you see the runway and determine that you are in a safe position to land, plan your missed approach as well as your landing. That decision point may come 1000 ft above the field on a beautiful day or 100 ft above the touchdown zone on an ILS, but until it does come, plan to take it around if the need arises. Brief the approach to a missed approach every time.

Salvage is a word used by amateurs. Don't put yourself in a position where you have to do everything perfect in order to get the aircraft on the ground. Instead of sweating and working to salvage a bad approach, take it around for another try. Get everything done and get stablized. It'll most likely be easier on you, your crew, and your passengers. The go-around and missed approach aren't just for when the weather has dropped below minimums, they're for when performance on this particular approach has dropped below a minimum safe level for any number of reasons. The missed approach is an ally in both situations, not an enemy.

Finally, even if you do everything right, something can still force you to go missed approach. We'll see in the next chapter how wake turbulence or windshear can do just that. If you are not prepared to fly the missed approach—you are not prepared to fly the approach at all. So review the missed approach procedure like your life depended on it, because often it does. It's your last chance.

9

Wake Turbulence and Windshear*

*This chapter was written by Noel Fulton.

We hear the warnings all the time. They've become part of the background noise of flying in an airport traffic pattern. "Cessna 37BY, winds 250 at 10. Cleared to land runway two four. Caution, wake turbulence." Or we hear the more ominous words "low level wind shear" on the ATIS frequency. We hear them all the time, but do these warnings and cautions always penetrate to change our behavior? Statistically, we know from the mishap and incident databases, that they do not, and that is the purpose of this chapter.

When's the last time you went around, delayed a turn to final, or slowed down on the basis of conscious considerations of wake turbulence or low-level wind shear? We all know the quiz answers about wake turbulence and windshear at groundspeed zero, but if we actually hit it in the air unprepared, it's final exam time, and the exam is graded pass/fail.

Windshear and wake turbulence are included together in this chapter because of their similarities. Both are radical disturbances of the air we fly through; one by Mother

Nature and one by our fellow aviators. When you learn about their characteristics and the standard techniques to avoid them, you'll strengthen the environment pillar in your personal airmanship model, which will better support and enhance situational awareness and judgment. The key to avoiding both of them is educated anticipation by pilots. You don't have to learn everything about windshear and wake turbulence to avoid them but if your knowledge is rusty, you could be in for a wild ride. Mild cases of windshear or wake turbulence may momentarily disturb your flight path. Serious cases can overwhelm the capabilities of your aircraft and threaten your life. In both windshear and wake turbulence, if it's serious enough, an immediate full power go-around is your best chance for survival close to the ground. We will start with a basic definition of windshear.

Windshear

Windshear is a change in wind speed and/or wind direction in a short distance resulting in a tearing or shearing effect. It can exist in a horizontal or vertical direction and occasionally in both. That definition is straight out of the Airman's Information Manual. Any number of factors can cause windshear. The most common causes of windshear are thunderstorms, passage of fronts near the airport, mechanical turbulence from buildings, and mountain wave. Temperature inversions can also cause shears. Windshear can be either performance enhancing or performance degrading. If you run into a 15-kn increase in wind on the nose, that will dramatically enhance the performance of your aircraft. Lose 15 kn, and performance will decrease. Both can be dangerous, as we saw in Chapter 5, "Performance Considerations."

An unexpected or drastic increase in performance can cause you to balloon on the glide slope. If no action is taken, the pilot is looking at a steeper-than-normal approach (if he or she keeps the same aim point) or a long landing. If the pilot reduces power to get back on the glide slope, a natural tendency, the pilot could be in trouble if he or she loses the extra headwind with the power back near idle. A decrease in performance can lead to a high sink rate that would be particularly dangerous low to the ground.

The key to windshear avoidance and recovery is to know that it exists—fundamental situational awareness. Equipment that detects dangerous shears is being improved all the time but is usually only available at larger, high-traffic airports where a low-level windshear warning will appear on the ATIS if detected. But if you aren't flying into a field equipped with this capability, there are still some clues available to possible shear activity. First, be aware of the local weather. If there are thunderstorms or fronts in the area, windshear is likely. Second, listen for or ask for PIREPS before you begin your approach. Asking specifically for any shear information may prompt another pilot to speak up. Third, consider calculating a reference ground speed for your approach.

Reference Ground Speeds

Most advanced aircraft have computers that calculate reference ground speeds automatically. The crew keys in the winds that are forecast or on the ATIS and the computer does the rest. If the navigation computer determines that the aircraft's actual ground speed is significantly more or less than the calculated reference ground speed, it alerts the cockpit crew, who takes appropriate action. Unfortunately, that equipment is expensive and the average GA pilot doesn't have it available. However, many GA aircraft do have basic ground speed readout associated with an onboard GPS or with relative speed toward a DME transmitter. The GPS ground speed will work regardless of the aircraft's track relative to a DME transmitter, but if all you have is a ground speed relative to the transmitter, the following technique will only work when you are flying directly to or away from a DME station on an approach.

This technique should sound pretty familiar. Many pilots commonly add a "gust factor" to their approach speeds. Some pilots add half the gust, some add it all. To calculate a quick reference ground speed, simply add the headwind or tailwind component of the forecast winds on the ground to your approach speed. TABLE 9-1 can help you out and is adapted from a table found in *Mental Math for Pilots* by Ronald D. McElroy (Cage Consulting, Inc., 2000). I'll use a similar table later when discussing crosswinds. Be advised the numbers are rough estimates only.

To use TABLE 9-1, first determine the angle between the runway and the forecast winds and choose the closest factor from the table. Next, multiply the wind by the factor to get a headwind or tailwind. Finally, add that component to your normal approach speed to determine a reference ground speed. Let's try a few problems assuming an approach speed of 120 kn (TABLE 9-2).

9-1 *Headwind/Tailwind Component*

Wind angle to runway	Calculated	Headwind/ tailwind	Component
0 or 180	1.0	100%	All
030 or 150	0.5	50%	Half
045 or 135	0.3	30%	One-third
060 or 120	0.1	10%	Almost none
090	0.0	0%	None

Once you have calculated a reference ground speed, you can use it to predict the type of shear you can expect and make a plan to penetrate it.

With your rough reference ground speed in mind, keep an eye on the ground speed indicated by your GPS or DME readout. Remember that many DME ground speeds are only accurate if you are flying directly to or from the station. Check the manual to make sure. On the approach, compare the actual ground speed with the reference ground speed you've just calculated. If your actual ground speed is higher than your reference ground speed; you can expect an increase in headwind or performance-enhancing shear somewhere on the approach. To penetrate a performance-enhancing shear, fly your approach speed as precisely as you can. It would be a mistake to try to fly slowly expecting more airspeed once you penetrate the shear. If visual, maintain a normal glide path. If you do balloon, get back on path as soon as possible and if the shear is too severe, take the aircraft around for another try. If you are expecting this shear on an ILS, maintain the glide slope as best you can. Do not attempt to fly below the glide slope expecting the shear to send you up to it. If the shear puts you too high to get stability back, take it around.

9-2 *Reference Ground Speeds*

Wind angle to runway	Total wind strength	Headwind or tailwind component	Reference ground speed
120	20 kn	10 kn tailwind	130 kn
050	20 kn	6 kn headwind	114 kn
140	30 kn	9 kn tailwind	129 kn
030	20 kn	10 kn headwind	110 kn
070	30 kn	3 kn headwind	117 kn

If your actual ground speed is lower than your reference ground speed, you can expect an increase in tailwind or performance degrading shear. To penetrate a performance degrading shear, increase your approach speed until your actual ground speed matches the reference ground speed you've calculated. This will give you enough energy to penetrate the shear without losing too much altitude or airspeed. It will also reduce the danger of a high sink rate or stall. Again, if the shear is too severe, add full power and go around. Finally, if you encounter severe windshear, give a PIREP. It is not just a good idea, it is a fundamental duty to your fellow pilots.

Give the PIREP as soon as you can, and be specific. Windshear is a temporary phenomenon and the longer you wait, the more it may increase, and more aircraft can fly into it. Try not use nonspecific terms like "positive," "negative," or "performance enhancing" in your PIREP. Leave those for the meteorologists. Pilots in the air can easily misunderstand them. For example, "negative windshear" could be interpreted as no windshear on the approach. Instead, try to describe the exact location and effect of the shear on your aircraft. For example, "Denver tower, Cessna November Foxtrot encountered windshear with a 25 kn loss at 500 ft on final for Runway 17L." If you had to take any drastic action, include that also: "A go-around with max thrust was required." You can also report any crosswind shears in the same manner.

Case Study: Any Landing You Can Walk Away from... (NTSB Report Number NYC97FA098)

"In May 1997, a Boeing 767-300, operated by Alitalia Airlines, was substantially damaged while landing at Newark International Airport (KEWR), Newark, New

Jersey. The 10 crewmembers, and 158 passengers were not injured. Visual meteorological conditions prevailed for the scheduled international passenger flight that was operated on an instrument flight rules (IFR) flight plan.

"According to air traffic control records, from the Federal Aviation Administration (FAA), the following Automatic Terminal Information Service (ATIS) was in effect at the time that flight arrived in the New York terminal area, and remained in effect through the time of the accident.

"'Newark tower information whiskey one seven five one zulu, wind three five zero at two four [knots], visibility one zero [miles], ceiling seven thousand broken, temperature one seven [Celsius], dewpoint one [Celsius], altimeter two niner niner four, ILS 4R approach in use, landing runway four right, departing runway four left...low Level windshear advisories are in effect....'

"The mishap crew made initial contact with the New York Terminal Radar Approach Control (TRACON) at 1412, and was told to expect an ILS Runway 4R approach, and was issued the current altimeter. About 10 seconds after the initial contact, the crew reported they had information WHISKEY, which was acknowledged by the controller.

"Examination of the air/ground voice communications tape from Newark Air Traffic Control Tower revealed that at 1427:55, and again at 1428:09, the local controller reported to arriving airplanes: '...gain or loss of fifteen knots below three hundred feet reported by several aircraft.' This information was not relayed to the mishap flight. ATC regulations don't require controllers to notify aircraft if the low level windshear warning is on the ATIS.

"The crew contacted Newark tower and was cleared to land. The captain later reported, 'We were cleared for the approach (ILS RNWY 04R); our indicated airspeed was 180 knots. We contacted the EWR tower and were cleared to land 04R, number five. Wind was reported 330/19/29. Our reference speed was 135 knots and, based on the reported winds, we adjusted our speed to a target of 150 knots approximately 7 miles from the airport.' This was predetermined by Alitalia's policy that calls for adding half the headwind component plus the gust. The minimum and maximum that can be added are 5 and 20 knots respectively. The headwind component was determined to be 5 knots with the wind 60 degrees off of runway centerline, plus the gust factor provided a plus 15 knots component to Vref. The captain stated, 'We cleared the fence stabilized in all respects on the approach. Well below 50 ft, AGL, the aircraft windshear detection system was activated. Within 2–3 seconds thereafter, the aircraft's left main landing gear touched down, followed by the right main and nose gear in that order. Given our altitude, speed, and power settings, we did not have an opportunity to react. We were, however, able to maintain directional control of the aircraft and prevent further damage to the aircraft and any injury to our passengers, who were safely evacuated via moveable steps.'"

Both crew members were experienced pilots; however, they only had a combined 218 hours in the 767. The captain had qualified in the aircraft about a month and a half previous to this incident.

Analysis

The warning of low-level windshear was "background noise" to this crew as it often is to many others. Two suggestions might have helped this crew avoid this incident. First, it is a good idea to review your personal aircraft or company emergency go-around or windshear recovery procedure every time you hear the ATIS report low-level windshear. A young captain I flew with did this every time. He was one of the most disciplined flyers with whom I've ever flown. We'd pull the Windshear Recovery Procedure out, read it, and think about it for a second during the approach brief whenever it came up on the ATIS. Second, if you hear the warning on the ATIS, but the controllers don't mention any PIREPS, ask for them. The Newark controller above told two previous aircraft about the, "gain or loss of fifteen kn below 300 ft," but was in the middle of switching shifts, so didn't tell the mishap crew. "Low-level windshear," is pretty vague. Try to get some more concrete and useful information.

Crosswind Shears

So far, we've just discussed windshear in terms of the increased headwind or tailwind and neglected the possible increase or decrease in the crosswind component. The primary danger in a crosswind shear is losing runway alignment. Sorry, no fancy techniques like "reference ground speed" here. The only way to guard against a crosswind shear is to know your aircraft's limits along with the crosswind component and develop a soft touch on the controls.

To help, I'll give you a shortcut to determine the crosswind component. TABLE 9-3 is taken directly from McElroy's *Mental Math for Pilots* and is the inverse of the one I used to determine the headwind component (TABLE 9-1). Instead of trying to memorize both tables, it may be easier for you to pick one and use it for both types of calculations. I would pick the headwind/tailwind table, because many checklists or in-flight guides have some sort of crosswind table already. Whichever technique or table you use, it pays to learn it well. Using one or both of the tables to quickly check if the crosswinds just gusted out of limits can save you from jeopardizing your certificate or your life. It can also help you anticipate how much crosswind control you can expect on final.

The Importance of Touch

The key to crosswind control on final is developing a soft touch on the controls. When I was instructing at the Air Force Academy, I used to run my new students through a common demonstration of trim in the airplane. I'd ask them to maintain straight and level as I slowly ran the trim wheel full up. Next, I'd ask them to give me a level turn. After the turn, I'd slowly move the trim wheel to the full down position and ask for another turn. The turns were usually pretty rough and pretty rough on the students who had to use a lot of muscle to keep the plane in a semblance of level flight. Once the second turn was completed, I let the students have the trim wheel to take the pressure off. Finally, they flew a third turn with neutral controls that was generally much more level and much less tiring on the students. One of the points of the demonstration was to demonstrate the difference between the fine motor control of their fingertips versus the gross control of the bigger muscles. I explained that good flying was like playing a musical

9-3 *Crosswind Component*

Wind angle to runway	Calculated	Crosswind	Component
0 or 180	0.0	0%	None
030 or 150	0.5	50%	Half
045 or 135	0.7	70%	Two-thirds
060 or 120	0.9	90%	Almost all
090	1.0	100%	All

instrument. Musicians use the small muscles of fingers and hands to tease out the melody and pilots had to use the small muscles to feel out the flight path.* The larger muscles are less sensitive, which lead to poor flying and the use of more energy, which leads to quicker fatigue.

That "musician's touch" is what helps pilots deal with crosswind shears. Make sure to trim the pressures off during your approach and continually monitor your alignment. Some pilots check their alignment and set their crosswind controls early on final. That is a good idea, but it's important to continually check the alignment, especially if there is a possible crosswind shear. Keep a light touch on the controls to help you feel the drift as soon as it begins and correct it. A light touch will also help you decide if the crosswind just became too much and it's time to go around. It doesn't matter if the aircraft in front of you just landed. That was then and this is now. Winds can change in an instant; don't let ego put you in a bad situation. There is an old saying that in aviation "flexible" is far too rigid. Real touch demands that a pilot stay "fluid." That wisdom certainly applies to crosswind shear landings.

Case Study: Almost Too Late (ASRS Accession Number 410887; Narrative by FO)

"After we were cleared for the approach to runway 30L, ATC advised us that a thunderstorm cell was on the field and there were a low level windshear and microburst alerts. We were flying in VMC and could see the leading edge of the cell and the rainshaft. I stated to the captain that, 'This is a textbook microburst scenario.'

---

*"Playing video games" substituted for instruments for my students who had never played an instrument.

"He was the pilot flying and elected to continue the approach and we heard aircraft Y say they were 50/50 on whether or not they would continue their approach. I asked ATC where aircraft Y was and they told me they were about three miles behind us. I said to the captain, 'Maybe we should go around,' and he replied that we should keep going.

"ATC came on the radio with another windshear and microburst alert reported by aircraft on the ground this time. The captain asked, 'When was that? Query ATC.' ATC replied that the report was 'current.' I again asked for a go around and the pilot responded that he was, 'going to keep it fast.'

"We encountered heavy rain as we entered the leading edge of the cell at approximately 500 feet AGL. I made the normal approach calls and was watching the airspeed closely. I did not notice any large fluctuations in airspeed but I did notice a very large crosswind and pronounced crab to the right of about forty-five degrees at 200–250 feet AGL. Airspeed was relatively steady at 130 knots and the crab was increasing as we drifted left of centerline. This time, we both called out, 'Go around!' We set max power, went around and called ATC.

"ATC offered us a vector to runway 4 and we accepted. During the return to runway 4, the pilot overbanked, descended and we received a, 'Terrain! Whoop! Whoop! Pull up!' warning. We were visual and the PF recovered. The visual approach to runway 4 was normal with a strong crosswind. Before the first flight of this day, the captain mentioned he only had three hours of sleep."

Analysis

This certainly seemed like a close call. First, to start with approach planning, the captain's lack of sleep should have been a red flag warning of possible poor judgment for both of these crew members. If the fatigue had been properly acknowledged and briefed, the FO may have been able to incorporate it into a more assertive statement at the beginning of this situation. For example, "Pilot, I am extremely uncomfortable shooting this approach into a thunderstorm that is reported on the field with your admitted fatigue. I think we should hold until the weather clears. What do you think?" From the narrative, the FO was obviously uncomfortable with the approach from the beginning yet was unable to get the captain to reconsider. It is incredibly hard to take an aircraft from a captain, but this situation almost certainly would have called for it if the captain hadn't finally initiated the go-around. One good thing this captain appeared to do was to carry some extra airspeed. It may have saved the aircraft and their lives but please, don't assume that a few extra knots will save you if you choose to shoot an approach into a thunderstorm. This crew was lucky. The next pilot may not be.

Closely related to windshear is the mechanical phenomena of wake turbulence, the subject of our next discussion.

Wake Turbulence

Wake turbulence is a part of flying. It rolls off the wings of every flying craft from ultralights to the Boeing 747. And while the wake turbulence from an ultralight is only a hazard to our feathered friends, the turbulence from the 747 is a hazard to everything smaller than it is. It's caused by the pressure differential between the high-pressure air on the underside of the wing and the low-pressure air on the top of the wing. This pressure differential produces two counterrotating horizontal "tornadoes" of air that trail behind the aircraft. Heavier aircraft can produce dramatic vortices that can exceed 300 ft/s, well beyond the roll authority of many aircraft.

Again, the key to avoiding this hazard is educated antic-ipation and avoidance by pilots.

Tests show that wake turbulence vortices usually descend from an aircraft more or less in line with the wings at several hundred feet per minute until they get within 100 to 200 ft from the ground. There, they start to spread outward along the ground at a speed of 2 to 3 kn. This assumes that winds are calm. Winds will move the vortices or cause them to stay in one position. The worst wind condition with respect to keeping at least one vor-tex in the same place on the ground is the infamous light quartering tailwind, which will center the vortex on the runway and blow it further down the field beyond the touchdown point of the aircraft that produced it. This is a particular hazard because pilots are trained to try and land beyond a preceding aircraft's touchdown point in order to avoid its wake turbulence. The Airman's Infor-mation Manual has some recommendations.

1. Landing behind larger aircraft—same runway: Stay at or above the larger aircraft's final approach flight path—note its touchdown point—land beyond it.

2. Landing behind a larger aircraft—when parallel runway is closer than 2500 ft: Consider possible drift to your runway. Stay at or above larger aircraft's final approach flight path—note its touchdown point.

3. Landing behind a larger aircraft crossing runway: Cross above the larger aircraft's flight path.

4. Landing behind a departing larger aircraft—same runway: Note the larger aircraft's rotation point— land well prior to rotation point.

5. Landing behind a larger aircraft crossing runway: Note the larger aircraft's rotation point—if past

the intersection—continue the approach—
land prior to the intersection. If a larger aircraft
rotates prior to the intersection, avoid flight path
below the larger aircraft's flight path or abandon
the approach.

6. Departing behind a larger aircraft: Note the larger
   aircraft's rotation point and rotate prior
   to the larger aircraft's rotation point. Continue
   climbing above the larger aircraft's flight path
   until turning clear.

7. Intersection takeoffs—same runway: Be alert to
   adjacent larger aircraft operation, particularly
   upwind of your runway.

8. Departing or landing after a larger aircraft executing
   a low approach, missed approach, or touch and go:
   Consider abandoning or delaying your takeoff or
   not landing until 2 min have passed.

9. En route VFR altitude: Avoid flight path below and
   behind large aircraft."

VMC and IMC wake avoidance

The above recommendations all assume that the preceding aircraft is clearly in view at all times. That is
because wake turbulence incidents are far more likely
to occur during VMC conditions when pilots are responsible for their separation. As soon as ATC clears an aircraft for the visual, the buck is passed and the pilot
assumes responsibility for separation. Basically, pilots
should be wary of any aircraft ahead of them that is the
same size or larger. If it's much larger, extra caution is
recommended in order to avoid an incident. Two minutes is usually the minimum separation you'd want
behind a larger aircraft unless you are positive you can
remain above its flight path. Flying upwind can also

help. When I was flying small turboprops into Denver International, I'd frequently get sequenced in behind heavy aircraft. If it were a visual approach I'd often set my aim point at the very end of the touchdown zone to stay above the danger. If you have a long runway, this is a viable option. If you don't, slow down and try to find that 2 min.

Wake turbulence is a little like alcohol. The only sure-fire remedy is time. You need time for the dangerous vortices to dissipate or move. All of the other recommendations about glide paths and aim points can help keep you safe but time between you and the other aircraft is the best safety net. If you can see the preceding aircraft well enough to determine when it crosses a point on the ground, you can try and hack the clock to check your spacing. This can be difficult in certain light conditions. If you can look for the aircraft's shadow moving across the surface, then that will be more precise than eyeballing a point in the sky over a landmark.

At controlled fields, the radios can help. Controllers generally hand aircraft off at about the same point. If you hack your clock when the aircraft ahead of you is sent to tower and assume you will be handed off at approximately the same point, this can also provide a clue to your spacing. Although this is very imprecise, it may be helpful. Aircraft with TCAS can sometimes determine a rough distance to the aircraft in front of them. That distance and your ground speed will help you find that 2 min. Finally, ask the controller what the spacing is. If the controller has time, he or she will be able to tell you. The AIM, paragraph 7-3-8-f-3, states, "During visual approaches pilots may ask ATC for updates on separation and ground speed with respect to heavier preceding aircraft, especially when there is any question of safe separation from wake turbulence." Not a bad idea.

Speaking of ATCs, they deal with pressure every day to cram more and more aircraft into an increasingly busy air traffic system. They have strict separation rules when the weather is IMC, but can bypass the rules when the weather is VMC by simply saying, "caution, wake turbulence," and passing responsibility to the aircrew. Generally this is perfectly safe because pilots can use the techniques already discussed. However, if a pilot feels that he or she can't maintain separation, the pilot needs to slow down and speak up to ensure that the safety of the aircraft is not jeopardized so ATC can keep to a schedule. Some pilots are even reluctant to call an aircraft in sight because that keeps the onus of separation on ATC's back. Occasionally, ATC will clear pilots for the visual while issuing a speed restriction such as, "maintain 170 to the marker." If you can fly the 170 *and* maintain comfortable spacing, fine. But realize that once ATC has cleared the pilot for the visual and made him or her responsible for the separation, the pilot can slow down if needed. ATC may complain but ATC is sitting on the ground and you and your passengers need that separation more than ATC needs the speed. If IMC, ATC will generally provide between 4 to 6 mi of separation between aircraft for wake avoidance depending on the relative weights of the aircraft. For example, a small aircraft can expect 5 mi of separation behind a 757 and 6 mi behind a heavy jet.* In IMC, pilots can provide an extra margin of safety with smart flying.

The same techniques a pilot would use to avoid wake turbulence on a visual approach will translate to an

*According to a study completed by the National Oceanic and Atmospheric Administration, the efficient design of the Boeing 757/767's wings generate, "vortex tangential velocities greater than any ever recorded, including those of the Boeing 747 and the Lockheed C-5A." During VMC conditions, treat them as you would the heaviest 747.

instrument approach. I am not recommending this technique, especially for pilots who are not experts and confident with instrument flying. However, I do want to make the reader aware that many professional pilots use it on occasion. First, if on a precision approach, fly slightly above the glide slope. A half a dot will be more than enough to keep you out of the wake turbulence of the aircraft in front of you (assuming it is on the glide slope). Next, take the winds into account and fly slightly upwind of the preceding aircraft. If the weather is above minimums, the pilot can use these techniques down to the runway by flying high and off center until the aircraft breaks out and then continuing visually down to the runway. Unfortunately, if the weather is right at minimums, the pilot will have to work back to centerline and glide path by decision height. Luckily, the pilot will still have the ATC minimum separation guidelines to keep out of wake turbulence.

For a nonprecision approach, plan to arrive at the minimum descent altitude near the visual descent point (VDP). The last thing you would want to do following a larger aircraft on a nonprecision approach would be to dive down to the MDA, level off, and drive toward the missed approach point. That will almost guarantee that you fly right into some wake turbulence. Remember, the key is to try and stay above the danger posed by heavier aircraft's vortices.

If you fly the heavies or know that a smaller aircraft is behind you on the approach, you can help a fellow airman by being predictable. The other guy is counting on you to fly the centerline of the course and the glide slope. That makes the other pilot's "educated anticipation" worthwhile. You or your flight computer could also easily plan a descent that would take you down to arrive at the VDP and MDA at the same time, but that would about guarantee that the person behind you would encounter your wake turbulence. If flying a nonprecision approach, it's not necessary for you to chop and drop down to the MDA and drive along it, but you can give your follower some room by planning your descent to arrive at your MDA a little before your descent point. A little courtesy could go a long way.

Case Studies: ATC versus Pilots

"We were being vectored for the ILS runway 27 at Boston. We were cleared the approach and told to maintain 170 knots until 5 DME. The controller put a 767 on the localizer about 5 miles in front of us and dropped him in from above our flight path. We ran into his wake, which violently rolled our aircraft. I clicked off the autopilot and started a climb to get out of the disturbance while slowing down for extra separation. The controller asked us our speed and we told him we had slowed to avoid more turbulence from the 767 and explained that we had just been hammered by the 767's wake. He definitely seemed more concerned that we were going to back things up on final." (ASRS Accession Number 477334; Narrative by Captain)

"We were flying our Brasilia EMB-120 on a fifteen mile final at 3500 feet when approach told us of traffic to follow was a MD88 3.5 miles ahead at 4000 ft. We had visual contact and were cleared the visual approach. I was not thrilled that the MD88 was 500 feet above us and would descend through our flight path so we slowed from 180 knots to 160 knots to allow more separation for the wake turbulence we expected to encounter. The autopilot was disengaged and as expected we encountered some turbulence at the glide slope intercept. The aircraft rolled rapidly left 30 degrees and then right 30 degrees.

"We recovered the aircraft with little difficulty and heard approach remark to an aircraft behind us that apparently we didn't want to obey his speed even though we had been cleared the visual!

"I was on a visual approach and had a large aircraft cleared in front of me from above with about three miles of separation. For the safety of the flight, I elected to increase the separation by slowing down to hopefully decrease the intensity of the expected encounter. I think that controllers should only put the same type of aircraft together if they want to send one aircraft through another's wake without standard separation. When I have been cleared for a visual approach, it is now my responsibility to maintain separation. When a larger aircraft is above my altitude on an approach and descending, I will slow down to increase separation. ATC had only three miles between us; I wanted at least four!" (ASRS Accession Number 487721; Narrative by Captain)

Analysis

The pilots in these studies acted correctly. They slowed to increase separation because they knew it was their responsibility. Maybe the first pilot should have slowed before encountering the wake, and slowing was definitely appropriate. The only other suggestion for avoiding the wake turbulence in both studies would have been for the pilots to fly slightly off course on the upwind side. If there are simultaneous visual approaches to closely parallel runways, this may not be an option, but it should be considered. You don't have to offset a great distance. Because vortices generally drop in line with the wings until they reach the ground, a hundred feet may be enough. Until you are sure you are above the previous aircraft's glide path, consider putting your wingtip on the upwind edge of the runway. If ATC challenges your reduced speed or course displacement, be polite but firm.

Both windshear and wake turbulence dangers are generally avoidable for the educated, situationally aware pilot. The keys to success and survival in this environment are simple, but must be done before an unanticipated encounter. Hopefully, the next time we hear the ATIS mention windshear or a controller cautions about wake turbulence, we can take preemptive avoidance or mitigation steps.

In the final chapter, some physiological and psychological keys to situational awareness will be examined, with an investigation into sensory issues and landing illusions.

10

Sensory Problems and Landing Illusions

Throughout the Controlling Pilot Error series, we have frequently pointed out pilot errors in terms of decisions and judgments. In this chapter, I will try to do something a bit different. We will look closely at *input factors*—those elements of information that form our perceptions of reality in the landing environment. Sometimes our perceptions are not reality, and follow-on decisions are doomed to failure. So our goal here is to unclutter the myriad of sensory inputs and try to patch up some common pilot errors related to misperception.

To accomplish these tasks, I will focus on those illusions that play out badly when they occur near the ground on approach and landing, those last few seconds when we have the opportunity to complete a successful flight—or turn it into a nightmare. Let's begin with the not so obvious notion that, in aviation, what you see is not always what you get.

Visual Illusions and the Landing Environment

In Chap. 7, "Midair Collision Avoidance," we discussed in some detail the workings of the eye with regard to the focal area and peripheral vision. You will recall that "foveal vision," accomplished with the most sensitive portion of our eyes, is extremely capable of discriminating between small differences. For example, foveal vision is used for reading, and also for scanning instruments, and looking for traffic conflicts. Peripheral vision, which is also known as "ambient" vision, is less discriminatory but sensitive to lines or large shapes, as well as relative movement. These differences are critical in the landing environment.

Our brain receives more sensory information from vision than it does from the senses of hearing, touch, taste, and smell combined. It is estimated that over 80 percent of our information about the world comes through vision. Visual illusions occur in two primary ways and they occur differently according to which part of our vision we are using—foveal or peripheral. In the first case, when using our foveal vision, we can misinterpret data, typically caused by brain processing problems with the visual data we are receiving from our eyes. For example, we look directly at a two-needle round dial airspeed indicator and misread it by 100 kn. This is not a far-fetched example; let me relate a personal story where this exact error occurred to me.

Many years ago, while I was a student in a small jet trainer, we were required to fly upwind leg (or "initial" in military jargon) at 200 kn. This was a good solid airspeed to integrate the arriving aircraft into the pattern flow as well as to support a 2G 60-degree bank "pitch out" to downwind. On this particular day, we were coming back into the traffic pattern in a hurry because the crosswinds were forecast to go out of limits and we needed to get some landing work in before that happened.

Proud of my efficient and smooth letdown into the traffic area, I found myself now getting behind the aircraft, making radio calls a few seconds late, and generally having a tough time getting the jet to trim out for smooth flight. On top of that, the aircraft in front of me was obviously flying well below the recommended pattern speed and I was going to have to extend my pattern to make up for this ineptitude. I had the throttles full forward, and couldn't understand why I wasn't able to get up to 200 kn. I began to be concerned that I might have the landing lights or partial speed brakes extended, so I did a quick configuration check. Nope—everything was retracted according to the checklist, no miscues for this intrepid aviator!

Glancing at my airspeed indicator, I *saw* 195 kn and unconsciously pushed a bit harder on the throttles, which were already at the forward stop. Wanting to impress my instructor with my understanding of the aircraft systems, I said "We need to write up these engines when we land. They just aren't performing like they should. It's probably a good thing though," I smugly continued, "because this dork in front of me is going even slower than I am." My instructor, now clearly amused, replied, "I'm not sure we really need to write up the motors, Tony. I've never been able to get *300 kn* out of them in the pattern either!" "Oh *no*!" I yelped, followed closely by an idle, full-speed brake pitch out to a misaligned downwind. To complete this unfortunately humbling story, I badly scrubbed a tire by landing in a crab, and after an ego-deflating debriefing I got some extra instruction on how the eye can play tricks on you. Here are some of the lessons I learned. In this episode,

I made a perceptual error with three primary underpinnings, often closely linked together.

1. *Expectancy.* We see what we expect to see. Because I was used to seeing 200 kn in the traffic pattern, I *perceived* 200 kn, even though the indicators were clearly showing 300. Another example of *expectancy* at work is a pilot who is used to seeing a 150-ft-wide runway will tend to adjust his glide slope to match his expectancy of what the runway should look like even if he or she is flying to a much wider or narrower runway. We will see examples of this in an upcoming case study.

2. *Ergonomic factors.* Sometimes the actual shape, location, and even color and contrast of an object can create perceptual problems. In this case, the indicator was not easy to read and could be misinterpreted by a low-experience pilot like myself.

3. *Overall visual acuity.* Issues such as ambient altitude, nutrition, and fatigue can affect our eyesight dramatically. If I recall correctly, my error occurred on my third flight of the day and I was not as sharp as I was earlier. (That's my story and I'm sticking to it.)

Peripheral Vision Problems

Our peripheral vision operates nearly subconsciously, providing a constant reference as to which way is up, what the horizon line looks like, our angle of bank or descent based upon perspectives provided in the outer portions of our visual field. This awareness occurs without effort or concentration, which is usually a good thing, but that can also work against us. For example, an insidious sloping cloud layer that becomes a false horizon might not be noticed without conscious effort. This brings us back to the need for constant scanning, giving your eyes an opportunity to focus and consciously read and react to what is now at the center of the visual field, as a means of validating what your peripheral vision has been telling you all along.

Because our peripheral vision operates at relatively low levels of concentration, it can be tricked, often by other sensory inputs that overpower ambient visual inputs to the brain. Sometimes this results in confusion, other times we are fooled completely. The brain uses a wide variety of wonderful inputs to shape its reality. These inputs include the eyes of course, but also the inner ears and skeletal muscle and nervous system. When all of the inputs agree, there isn't a problem, but when they don't, there can be problems.

The "Eyes" Have It, or Do They?

Let's start with the obvious. Things look and feel different in the air as opposed to our daily *terra firma* existence. Objects appear smaller and are viewed from a completely different angle than we are accustomed to viewing them. They move in our field of vision more quickly, and ironically, those objects that do not move might be coming at us faster than those that do! Lights on the ground may appear to be in the sky or vice versa. Distances are difficult to judge. Large trees may be mistaken for small shrubs or, worse, small shrubs may be mistaken for large trees, leading pilots to *assume* (the mother of all screwups) that they are higher than they actually are. In short, spatial information processed by the human eye tuned for ground observations, are potentially unreliable, and more so in the less experienced pilot.

The Skeletal-Muscular System

For perhaps millions of years, humans have grown accustomed to the comfortable 1 g pull of gravity. This creates a certain "feel" or tension in the muscular system that is trusted almost without question by earth dwellers. This creates a couple of problems for aviators. First, we know from our basic aerodynamics (as well as from roller coaster rides) that acceleration forces provide a feel of artificial gravity. When these are linear, we feel simply heavier, pushed back into our seat, but when we combine these straight-line forces with rolling moments, deceleration, climbs, and dives—and when the inner ear kicks in—then this spatial awareness thing gets interesting, and terms such as "up" and "down" need verifying.

The Inner Ear for Dummies (a.k.a. Pilots Like Me)

I don't know why but I've always liked the term *semicircular canals*. It just sounds cool—and they really are cool, when they work to our benefit. When they don't, it results in a pretty interesting ride, as anyone who has suffered through a bout of the *coriolis illusion* and lived to tell about it can readily attest. To keep this discussion at a level I can understand, and therefore write about, I am going to omit terms like *ossicles, cochlea, ampulla, macula,* and *endolymph fluid.* Instead, I will try and simplify this physiological wonder as fluid inside a tube, moving over sensitive hairs to give you a sense of motion and spatial awareness.

Dr. Richard Reinhart, the author of *Fit to Fly: A Pilot's Guide to Health and Safety* (McGraw-Hill, 1993) uses the analogy of a field of grass gently swaying in the breeze to describe how the inner ear works. I prefer the analogy of a lover blowing gently across the back of your neck. In either case, what we are sensing is relative motion. Several millennia of evolutionary change have superbly prepared our bodies to accurately sense relative position, using the inertia of the fluid across the sensitive hairs to keep us in a state of equilibrium— where perceptions and reality are the same. But airborne forces like acceleration and deceleration often cause a false sense of motion, resulting in several types of spatial disorientation.

As you can see, many factors are involved with disorientation and illusions, some are internal as we have just discussed and some are external, such as the optical characteristics of a windshield, rain, fog, haze, glare, or dust. Additionally, the angle of approach and contrast and slope of terrain can also set us up for an unwelcome surprise. Well, enough of the aero medical background information for now. Let's move on to a discussion of the common illusions that can lead to fatal pilot errors.

Common Visual Illusions

The U.S. Army likes to fly close to the ground, and has spent a great deal of time and effort identifying common illusions that lead to mishaps in this environment. They have identified the following 11 illusions that are taught to all new pilots. All can affect us negatively in the approach and landing environment.

1. *False horizons.* Cloud formations may be confused with the horizon or the ground. Momentary confusion may result when the aviator looks up after having given prolonged

attention to a task in the cockpit. Because outside references for attitude are less obvious and reliable at night, aviators should rely less on them in night flight. A related illusion is the disorientation caused by the northern lights or city lights reflecting off low clouds.

2. *Flicker vertigo.*   Much time and research have been devoted to the study of flicker vertigo. Lights that flicker at a rate of 4 to 20 cycles per second can cause nausea and vomiting; in severe cases of flicker vertigo, convulsions and unconsciousness may be experienced. Fatigue, frustration, and boredom tend to add to these unpleasant and dangerous reactions. Although this rarely occurs, flicker vertigo must be recognized as a potential problem. Flicker vertigo can be caused by sunlight flickering through rotor blades or propellers or by strobes or rotating beacons reflecting against an overcast sky.

3. *Fascination (fixation).*   Fascination occurs when aircrew members ignore orientation cues and focus their attention on their object or goal. Target hypnosis is a common type of fascination. For example, a pilot may become so intent on hitting the target during a gunnery run that the pull-up is delayed too long and the aircraft contacts the ground. The same thing can occur to a pilot that stares too intently at his landing aim point.

4. *Confusion with ground lights (ground light misinterpretation).*   At night, many pilots have put their aircraft into very unusual attitudes to keep ground lights *above* them; they mistook

ground lights for stars. Less frequent, but just as dangerous, are the illusions caused by certain patterns of ground lights imagined to be things that are not actually present. For example, some pilots have mistaken seashore lights for the horizon and, believing they were flying straight and level, have maneuvered their aircraft dangerously close to the sea. Some aviators have also confused certain geometric patterns of ground lights, such as moving trains, with runways and approach lights and have been badly shaken by their near misses.

5. *Relative motion.*   Relative motion is best explained by using an example. A person sits in a car and waits for a stoplight to change, and another car is driven up alongside. As the other car is picked up in the peripheral vision, the person senses movement in the opposite direction. This illusion is often encountered during formation flight when pilots see the movement of another aircraft and interpret it as their own motion.

6. *Autokinetic illusion.*   The autokinetic illusion results when a static light appears to move when it is stared at for several seconds in the dark. A person can see an example of this illusion by taking a lighted cigarette into a completely dark room and staring at it. Apparent movement may begin after 6 to 12 s. Uncontrolled eye movement may possibly cause the illusion of movement as the eye attempts to find some other visual reference point. (a) The autokinetic illusion can be reduced or eliminated by visual scanning, by increasing the number of

lights, or by varying the light intensity. (b) The autokinetic illusion is not limited to lights in darkness. It can occur whenever a small, bright, still object is stared at against a dull, dark, or nondescript background. Similarly, the illusion can occur when a small, dark, still object is viewed against a light, structureless environment. The observer may experience this illusion any time there are no visual references. To ensure safe operations at night, aviators need to be aware of this condition and know how to cope with it.

7. *Structural illusions.* Structural illusions are caused by heat waves, rain, snow, sleet, or other visual obscurants. A straight line may appear curved when viewed through the heat waves of the desert; a single wing-tip light may appear as a double light or in a different location when viewed during a rain shower. The curvature of the aircraft windscreen can also cause structural illusions. These illusions are due mainly to the refraction of light rays as they pass through some visual obstruction.

8. *Height perception illusion.* When flying over desert, snow, or water, pilots may experience the illusion that they are higher above the terrain than they actually are. This is due to the lack of visual references. Also, pilots that are used to flying over areas of large trees may interpret small shrubs falsely and fly too close to the ground. Flight in an area where visibility is restricted by haze, smoke, or fog produces the same illusion.

9. *Size-distance illusion.* The size-distance illusion results when an observer stares at a point of light that appears to approach and recede. In the absence of additional distance cues, accurate range estimation is extremely difficult. Instead of seeing the light approaching or receding, aviators may see the light as expanding and contracting at a fixed distance. They can dispel this illusion by using proper scanning techniques.

10. *Altered planes of reference.* When aviators approach a line of mountains or clouds, the planes of reference are altered. They may feel that they need to climb even though their altitude is adequate. When flying parallel to a line of clouds, they may tend to tilt away from the clouds.

11. *Reversible perspective illusion.* At night, an aircraft may appear to be going away when it is actually approaching. This illusion is often experienced when an aircrew member observes another aircraft flying a parallel course. To determine the direction of flight, the aircrew member should observe the aircraft lights and their relative position to the horizon. If the intensity of the lights increases, the aircraft is approaching; if the intensity dims, the aircraft is going away.

Each of these visual illusions is hazardous and potentially lethal, and exist in normal flight operations as well as in the approach and landing environment. The following illusions are associated with specific aspects of the landing environment.

Seeing Things from a Different Angle

Earlier in this chapter we talked briefly about expectancy or seeing what you are used to seeing, even if the situation is different. This can be a very difficult challenge for a pilot of low or singular experience, who operates primarily out of a fixed base or out of the same runway environments as a matter of routine. Let's look at a few of the more common illusions associated with these phenomena and analyze one with a case study.

- *Upsloping runway or terrain.* This environment can give the pilot the illusion of being too high or too steep on approach, often leading to a push over or a less-than-desirable glide path angle. Danger: Landing short or impacting terrain.

- *Downsloping runway.* Conversely, a runway that slopes away from the pilot may appear to the pilot to be a picture of a shallow approach path, causing a pitch up and land long. Danger: Long landing and inadequate runway for stopping.

- *Narrower-than-usual runway.* Similar to the upsloping runway illusion, this illusion can make a pilot feel as if the aircraft is higher than it actually is, as the brain tries to make sense of a different picture. The result is a tendency for the pilot to push over and fly a lower approach in an attempt to make the picture appear "normal." Danger: Landing short or impacting terrain.

- *Wider-than-usual runway.* A pilot used to landing on a 50-ft-wide runway will be quite surprised by the appearance of a 150-ft- or 300-ft-wide runway on the same 3-degree glide slope. The tendency is to feel low and to climb to make the

> picture match our expectations. Danger: Long
> landing and inadequate runway for stopping.

Let's show how this plays out in the real world, with an example from an accident in Wisconsin, where a low-time student pilot fell victim to a classic landing illusion.

That Looks about Right (NTSB Report Number CHI98LA061)

"On December 13, 1997, at 1545 central standard time (CST), a Piper PA-28-140, N5454S, piloted by a student pilot, was destroyed during a collision with a moving tractor-trailer truck and terrain while on short final approach to runway 26L (2,300 ft × 33 ft dry/asphalt) at the Sylvania Airport, Sturtevant, Wisconsin. Visual meteorological conditions prevailed at the time of the accident. The pilot was fatally injured. The flight departed Sturtevant, Wisconsin, at 1540 CST.

"Witnesses said N5454S's main landing gear struck the semi-trailer truck's trailer. They said the airplane was approaching the runway from the east '...at a very low angle,' according to the Racine County Wisconsin Sheriff's Department report. One witness said the airplane '...had a slight up and down jerky movement as it crossed the northbound traffic just before hitting the semi [that was driving in the southbound lane].' Another witness said that the '...airplane was flying low toward the ground and as it got closer it looked as if it hit turbulence because it sort of swerved lower and hit the top right of a semi-truck. Then its wing skidded off the truck....'

"Witnesses reported that the airplane collided with the ground in a pitch down attitude with the right wing lower than the left wing. They said the airplane rebounded and slid backward about 20-feet from its impact point. The initial ground collision point was about 100-feet west of the highway where the truck was operating.

"The pilot was a student pilot who began her flight training on April 26, 1997. At the time of the accident, the pilot's logbook showed she had 37.3 hours of flight time. The pilot's logbook showed she had 26.6-hours of flight instruction and no solo flight time in a Cessna 150 before transitioning to the Piper PA-28-150 (PA-28). She had 1.6-hours of solo flight time in the PA-28. The pilot had 9.1 hours of dual instruction in the PA-28 series airplanes and soloed the PA-28 after she had received 3.1 hours of flight instruction. The logbook showed she had received instruction in the following areas and sequence: 'Touch-goes, stalls, aircraft system orientation (.9-hour flight instruction), crosswinds [takeoff and landings], 190/18 with gusts, go-around (.7-hour flight instruction), simulated power failure with landing, landings and take-offs—looking good (1.0-hour flight instruction), 5 dual [takeoff and landings], 5 solo landings (1.0-hour total time, .5-hour solo).'

"Weather reported at the Kenosha Municipal Airport, Kenosha, Wisconsin, on December 13, 1997, at 1550 central standard time were a clear sky, visibility of 10 statute miles with winds from 260-degrees magnetic at 6 knots.

"The accident airport is privately owned but open to public use. The State of Wisconsin has designated the airport as a reliever airport for general aviation aircraft using Mitchell International Airport located in Milwaukee, Wisconsin. The Milwaukee, Wisconsin, Federal Aviation Administration Flight Standards District Office Principal Operations Inspector said his office has had complaints of low flying airplanes by drivers on the interstate highway that is about 300-feet from the runway's threshold.

"The displaced threshold begins about 160-feet from the interstate highway. During subsequent investigation activity, the IIC observed airplanes flying about 20 to 50 feet above the vehicles that were using the highway as the airplanes approached to land. Some airplanes landed on the displaced threshold. The displaced threshold markings on runway 26L were weathered and very light in color.

"The pilot had been flying at Mitchell International Airport, Milwaukee, Wisconsin. The runways most often used at this airport for flight training were 3,163 ft × 100 ft and 4,182 ft × 150 ft. The runway at Sylvania Airport is 2,300 ft × 33 ft.

"According to the textbook *Human Factors in Flight* by Frank Hawkins, runway width differences may be a source of distortion in perception during landing. When the runway is narrower than normal the pilot may experience an illusion that causes them to fly a lower than normal final approach path.

"A human performance specialist from the FAA's Civil Aeromedical Institute, AAM-510, who is part of the Flight Crew Performance section, stated that this happens because the pilot is trying to match the mental site picture of the wider runway to the narrow runway. By flying lower the site picture begins to resemble the wider runway. The book, entitled *Flightdeck Performance: The Human Factor,* by David O'Hare and Stanley Roscoe, states, 'Some researchers have suggested that if the width and/or length of an unfamiliar runway differs radically from that to which the pilot is accustomed, then the resulting illusion may cause systematic deviation above or below the desired glide path.'"

The National Transportation Safety Board determines the probable cause(s) of this accident as follows: "The pilot's misjudgment of altitude/distance and her failure to maintain adequate visual lookout. Also causal was the pilot's failure to maintain adequate obstacle clearance. A contributing factor was the pilot experiencing a visual illusion."

There are a couple of lessons here. First is the obvious implication of flying to a field that looks much different than your "home 'drome." The illusion caused by the narrower-than-usual runway resulted in the classic "drug in" approach. The second lesson is one about experience—or lack thereof. In seems in this case that many aircraft were having difficulty flying into this field. How much more likely is it for someone who may have *never seen* a short and narrow runway to have a mishap? The local aviation community used this mishap as a springboard to recommend glide path indicators (VASI) for this field.

This brings up an interesting point. So far we have discussed common errors, but are there also common *locations* where these errors tend to manifest themselves disproportionately? Are airfields in your area especially challenging with regard to the potential for visual illusions? Perhaps most important, how do you protect yourself from falling victim to this dangerous combination? The next two short case studies illustrate such a scenario.

Island Time and Again (NTSB Report Number LAX96LA136)

"On March 9, 1996, at 1300 Pacific standard time, a Cessna 182Q, N7583S, sustained substantial airframe damage during a hard landing at the Catalina Island Airport, Avalon, California. The aircraft was rented by the pilot for a personal flight. Visual meteorological conditions prevailed at the time and no flight plan was filed. The private pilot and his two passengers were not injured. The flight originated at the Whitman Airport,

Pacoima, on the day of the accident at 1215 as a personal flight to Catalina Island.

"In a telephone interview, the pilot reported that this was his first flight to the island and he was unfamiliar with the airport. The airport sits on a mountain top with steep cliffs on both the approach and departure ends of runway 22. [See FIG. 10-1.] The first 2000 feet of the 3240-foot-long runway has a pronounced upslope gradient. The pilot said he made three attempts to land and was high and fast each time due to the optical illusion presented by the upslope runway. On the third approach, the aircraft touched down hard and bounced back into the air. The pilot said he then released back pressure on the control yoke and the aircraft hit hard on the nose wheel, damaging the propeller and firewall."

The National Transportation Safety Board determined the probable cause(s) of this accident as follows: "...the pilot's improper flare and improper recovery from a bounced landing. Factors relating to the accident were: a visual illusion due to rising terrain at the approach end of the runway, and the pilot's misjudgment of distance and speed during the approach to land."

In this event, the combination of unfamiliarity and the upsloping terrain proved too much for this pilot. In the next example, taken from the same airfield, the damage was not limited to the pilot's own aircraft.

Catalina 2 (NTSB Report Number LAX96LA235)

"On June 16, 1996, at 1502 hours Pacific daylight time, a Cessna 172N, N6261D, collided with three parked aircraft after a hard landing at Catalina airport, Avalon, California. The aircraft sustained substantial damage; however, neither the private pilot nor his three passengers were

10-1 *Catalina Airport. Sharp terrain drop-offs on each end of the field make landing illusions a serious challenge at Catalina.*

injured. The flight originated in Santa Ana, California, about 1430 on the day of the accident. Visual meteorological conditions prevailed at the time and no flight plan had been filed.

"The airport manager reported that he first observed the aircraft in the traffic pattern on approach to runway 22. During the landing sequence the aircraft failed to flare and touched down on the nose wheel. The aircraft porpoised twice and on the second bounce, the pilot attempted to recover. The aircraft veered off the runway and stalled about 25 to 30 feet in the air, hit the ground, and subsequently struck three other parked aircraft.

"The pilot reported that he was making a normal approach with 40 degrees of flaps. On touchdown he bounced once and then bounced a second time, when the aircraft struck a hole in the runway. The nosewheel began to shake severely and he applied back pressure on the control wheel but thought at this time he was nearing the end of the runway. He intentionally turned left to avoid running off what he thought was the departure end.

"A second witness, who was a coworker of the airport manager, reported that he heard the engine accelerate during the accident sequence. He also stated that there were no holes in the runway at the time of the accident.

"According to the NOAA Airport/Facility Directory, pilots cannot see aircraft on opposite ends of the runway due to the runway gradient. The other aircraft involved were identified as: a Beech G18S, N9375Y; a Grumman AA-5, N5439L; and another Cessna 172N, N6441D. A post accident inspection of the aircraft revealed that the flaps were still extended to 40 degrees and the aircraft was trimmed nose down."

In each of these cases, the pilots were confronted with visual illusions that are common knowledge among pilots who fly regularly into Catalina, and readily available to those who don't. The mishaps were caused by illusions created by known terrain and runway features. The corrective actions here are relatively simple.

1. Familarize yourself with the terrain and runway environment you are flying into. There are many methods for accomplishing this critical step, including reviewing the information available in various commercial or government documents, such as the NOAA Airport/Facility Directory, or talk to pilots who are familiar with the field. This may require a phone call, but it is well worth the effort. Many airfields now have websites filled with valuable information and photographs. I found the

picture of Catalina Airport I used here on one such website.

2. Prepare yourself for the potential landing illusion. Read this chapter, understand what sloping terrain might do to your perception of the situation and develop a countermeasure. Perhaps you will bring your altimeter and your vertical speed indicator into your cross-check a bit more, or perhaps you can brief your passenger to provide altitude calls for the last 100 ft prior to touchdown. If you are a relatively inexperienced pilot, it might be a good idea to find an instructor who can take you to some local fields that present landing illusion challenges, so you can gain a bit of proficiency. But usually awareness—and the ego control to take it around if it doesn't look right—is enough of a countermeasure for most experienced pilots.

Our next topic deals with a common setup for a sensory illusion that has nothing to do with the physical features of the landing environment, but is still responsible for numerous accidents and close calls.

Transitioning between Instrument and Visual Conditions (and Back Again)

At first it seems pretty straightforward. You are flying an instrument approach because you are in the weather, at some altitude prior to your minimums, and you hope to break out, at which time you will transition to visual flight rules and land. What could be simpler? As with many things in life that appear simple, the devil is in the details. Transition flight may be one of the most difficult things that we do, and not just from IMC to VMC. There is an equal set of challenges going in the other direction, when we are flying an approach in decent weather and suddenly find ourselves inside a cotton ball with a need to transition to instruments. Even for trained and proficient instrument pilots, there are times when our physical senses are fooled, creating a dangerous condition. Let's take a moment to look at each case.

Breaking Out: IMC to VMC

Transitioning from complete instrument flight (where the pilot has an internal scan methodically working to match a known set of desired parameters to what they are actually seeing by analyzing trends on a small set of instruments) to visual flight (where the pilot should be doing the exact same thing with the outside world as one of the "instruments") bites far too many aviators on approach and landing. There are two classic errors that we must discuss, and although neither is a "landing illusion" in the purest sense of the word, they are tied so closely to sensory inputs and expectancy factors that I feel the need to mention them in this chapter, if only briefly.

The "Duck Under"

When pilots begin an instrument approach, they generally expect to break out if the weather is reported above minimums. They trust their skills and their aircraft, and know if they ply their craft correctly, the earth will appear pretty much on schedule. What we tend to forget on occasion is that the weather can often change from moment to moment; our instrumentation may be operating within acceptable tolerances, but still be slightly off; and when coupled with a lack of proficiency or slow cross-check—well, you know the drill (and if you don't, reread Chap. 8, "Missed Approach")—the runway doesn't appear on schedule or in the place where you expect it.

As we approach our landing minimums and the earth begins to appear in bits and pieces directly below our aircraft, many of us feel the compulsion to nose the aircraft over to see more of what we want and expect to see. The result, of course, is an aircraft dropping below a safe glide path, and if not corrected quickly, can lead to big trouble in a hurry.

We counter this tendency, like all visual illusions, through knowledge, discipline, and procedure. The knowledge we need is here, as well as in multiple NTSB case files. You must know that the duck under maneuver is inherently dangerous and can kill you. The disciplined use of technique and a standard operating procedure can prevent the duck under maneuver. Most crewed aircraft in commercial airlines and the military teach two remedies for the duck under. The first technique is to ignore ground sightings directly below your aircraft and to transition to outside references *along your projected flight* path at the same rate that they become available to you. This allows the pilot to use a composite cross-check to transition gradually, finally using primarily visual cues with an instrument backup in the final phases of the approach.

In a crewed aircraft this often translates into a standard operating procedure where the pilot not flying (PNF) makes specific callouts, such as "I have the approach lights—I have the runway environment—we are visual." This is done until there are sufficient cues to land allowing the pilot flying to fly pure instruments. When the pilot flying transitions outside the cockpit for a visual landing, the PNF comes "inside" to monitor the instruments for any potentially hazardous change (like a duck under) that might have resulted from the transition.

With a good briefing to a knowledgeable passenger or fellow pilot, general aviation pilots can use the same tools. Occasionally, pilots are too quick to "go visual" and can create a whole nest of problems that can range from embarrassing things, like landing at the wrong airport, to deadly things, like slamming into terrain.

"Report the Runway in Sight": A Setup for the Visual Trap

It happens almost every time we approach an airport under ATC control, the controller will say something like "(Call sign), descend and maintain four thousand, fly heading 170 and report the airport in sight." You adjust your altitude and heading and begin to gaze toward the spot where the field should momentarily appear, eager to show the controller that you know where you are at and where you are going. But what comes next can determine the safe completion of your flight, or set you up for trouble. Because once you report the runway in sight, the odds are that the controller is going to offer you a visual approach, and if you accept it, you have accepted responsibility for separation and navigation. You may have complicated your approach and possibly increased your workload.

Unless you are very familiar with the airfield and the local environment, and can accept a wide variety of visual landing instructions from the tower controller, it might be a good idea to keep your instrument clearance and fly the published approach. Many, many wrong airport landings have resulted from exactly this situation.

The final topic for this section deals with the challenge of *reentering IMC* conditions after we are on a visual approach, and this can create another hazard, one closely associated with a classic and potentially lethal sensory illusion.

Back In: Visual-to-Instrument Transitions on Approach

Just as we need a disciplined procedure for breaking out of the weather on final approach, we must have a plan for reencountering IMC conditions. Typically this occurs as a result of two situations, slant-range visibility problems and ground fog. Each presents a different challenge.

Slant-Range Visibility Challenges

Slant-range visibility is the visibility along your projected flight path or from your altitude to the ground along the angle of your approach. Visibility readings taken at the airfield are measured at the surface, and are usually predictive of what a pilot will have available on approach, but occasionally atmospheric conditions (dust, smoke, fog, haze, diffused light, etc.) are such that the slant-range visibility is far less than they are reporting at the field. When this occurs, we are set up for the same problems that arise from going visual too early when we begin to pick up glimpses of the ground beneath our aircraft on a normal "through the weather" approach. The inclination to "duck under" where we can see a bit better is even more prevalent under conditions of poor slant-range visibility for two reasons. First, we are often already VFR and therefore operating legally under "see and avoid." That 200-ft radio tower might not be visible under these conditions, so once again, unless you are extremely familiar with the local area and all the hazards, it might be best to stay at the instrument minimums until you have the landing environment clearly in sight.

This doesn't always occur on approaches to well-marked fields, as our next case study illustrates.

Case Study: A Creative Approach (NTSB Report Number CHI98LA083)

"On January 20, 1998, at 1107 central standard time (CST), a Nerstrom Tailwind W-8, N12038, piloted by a private pilot, was destroyed during a collision with the frozen, snow covered surface of Lake Wisconsin during a low pass. Visual meteorological conditions prevailed at the time of the accident. The 14 CFR Part 91 personal flight was not operating on a flight plan. The pilot reported no injuries. The flight departed Baraboo, Wisconsin, at 1030 CST.

"The pilot stated in his written statement that he left Baraboo, Wisconsin, for a short local flight. He said that after 30 minutes of flying he descended over Lake Wisconsin, near Lodi, Wisconsin, to check and see '...if anyone had been using the lake at [the] restaurant area...' for landing. He said that he 'dropped down to abut 50 feet over the lake [while] looking out the left door window...' at his shadow. When the airplane went under a cloud, the pilot said he realized he could no longer see his shadow. The pilot said he contacted the ice '...the instant [he] could no longer see [his] shadows.'

"During a telephone conversation, the pilot said he lost visual reference when the airplane's shadow disappeared. The pilot said that he could not tell where the lake surface was due to the slant range visibility and whiteout conditions he encountered. He stated he pulled up to gain altitude, but the main landing gear contacted the ice, shearing it off. He said on the subsequent bounce, the wing struts were overloaded and the wings broke off."

In Flight Training Manual by Transport Canada, overcast whiteout is described as "a uniform layer of cloud over a snow-covered surface." The sunlight is "scattered and diffused" in between the overcast clouds and snow surface in all directions. This results in "the space between the ground and cloud appear[ing] to be filled with a diffused light with a uniform white glow. Depth perception is completely lacking as the sky blends imperceptibly with the ground at the horizon line, causing disorientation."

This pilot comes up with a creative—if temporary—solution to the problem of poor slant-range visibility, by using his aircraft shadow for ground reference. Are you ready for the punchline?

Because the pilot lost sight of his shadow in this mishap (which happened only a few days away from Groundhog Day) all of Wisconsin had to suffer through six more weeks of winter.

Ground Fog Presents a Different Challenge

Unlike the problems associated with slant-range visibility, ground fog encounters usually occur suddenly and very close to the ground. The actions you take are going to depend on several factors, including the height at which you encounter the fog, the stability of your approach, and the relative position of your aircraft in relation to the runway. A serious ground fog encounter is one of those rare times in aviation where you must react quickly and positively to avoid a mishap. In the next case study, the pilot flying an approach encountered ground fog conditions and made a series of questionable decisions that ended in tragedy.

Case Study: Ground Fog Encounter (NTSB Report Number SEA99FA027)

"While the pilot was en route to his intended destination on a dark night, the weather lowered to less than visual flight rules (VFR) minimums. Approach control therefore cleared the pilot direct to the associated VOR for a VOR DME approach. Once at the VOR, the pilot said he could see the ground and was going to descent and make a visual approach. Approach control asked the pilot if he was asking for a contact approach or canceling his IFR clearance, and then explained to him that he could not be cleared for a visual approach unless he could see the airport. The pilot responded that he wanted to shoot a contact approach, and he was cleared to do so. Soon thereafter the aircraft was seen maneuvering underneath a low ceiling, in hilly terrain with ground fog present. While maneuvering in the hilly terrain, the aircraft collided with a row of pine trees near the top of a hill, and continued on approximately 1000 feet before impacting a thick stand of hardwood trees killing the pilot and three passengers."

The National Transportation Safety Board determined the probable cause of this accident as follows: "[T]he pilot's failure to maintain clearance from a hilly terrain while maneuvering below a low ceiling. Factors include darkness, low ceiling, ground fog and hilly terrain."

Once again we see a pilot faced with a challenge in the landing environment, making a series of questionable judgments that lead to a mishap. Perhaps this is the best lesson of all, that when things begin to unravel at or near the end of the flight, the potential for a rapid error chain is always present.

The Time Trap

Although we know from theoretical studies that time plods along at a steady pace (at least, according to Einstein, on this side of the speed of light squared), we also know from our experience that time appears to accelerate as we draw near the end of a flight. This is because events become more tightly coupled and more sequential—or linear—in the approach and landing phase of flight. At altitude, during cruise, we have the luxury of reviewing and double-checking systems and other factors associated with a safe and efficient flight. Once we begin our descent, however, things come more quickly—checklists, communications, approach reviews, weather decisions, and so forth. A single error impacts not only the next event or decision—but often it limits the time available to accurately assess the new situation.

If I can get you to recognize and accept this one fundamental truth—that *time* is your biggest enemy or strongest ally in the approach and landing environment—it will have been worth the effort to write this book. In several of the case studies we have seen in this book, that fundamental truth was forgotten, and pilots took actions that increased their workload or shortened the time available to deal with a changing situation. Accepting a visual approach when you may not be fully prepared is only one of many ways that pilots do this to themselves.

To assist in helping address this most fundamental issue, Perry Thomas penned a checklist to assist pilots in preparing for the challenge of change in the approach and landing environment. It is taken from an *ASRS Directline* and reprinted with permission.

One of our better ways of learning how to stay alive in the flying game is to profit from the unhappy experiences of others. Here are a few practical, no-nonsense suggestions from these same reports that should reduce some of the hazards.

For all approaches

- Review and brief all applicable visual and instrument charts before the approach.

- DO NOT identify traffic in sight, airport in sight, or runway in sight, unless you are certain of your identification, and your flight deck (or cockpit) mates concur.

- Keep your traffic in sight; if you lose your traffic, tell ATC.

- Ensure that at least one pilot monitors the gauges and radios to "aviate, navigate, and communicate."

- Use all available electronic navigation to back up a visual, if you do decide to accept it.

- If visual approaches are being conducted but you don't want a visual, insist on an ILS or other instrument approach. Bear in mind, however, that during your instrument approach, other aircraft in your proximity may be conducting a visual approach.

- Expect visibility to deteriorate and be reduced if you are descending into a smog/haze layer (and possibly the setting or rising sun), during the turn to base and final. This may lead you to misidentify the runway to which you are cleared.

For parallel approaches

- Be aware that parallel runway approaches means that there is likely to be other traffic close at hand. There may be a significant increase in flight deck workload—unless the flight crew members brief and prepare themselves to the maximum extent possible. Safety in visuals will be enhanced by close coordination between flight crew members, and by maintaining a careful traffic watch outside the aircraft.

- Beware of overshooting runway alignment and encroaching into the parallel runway's approach path.

- Beware the dangers of "the visual trap."

A Final Word on Landing Environment Error and Sensory Illusions

So what can pilots do to improve their odds? You are already taking the first step; learn about the challenges and be able to recognize deteriorating conditions—your own, the aircraft's, and the environment's.

The second key skill is to *anticipate.* Think ahead, play the "what-if?" game. Learn about the different illusory phenomena we have touched on here and understand what triggers them in you. Stay aware of changing conditions and your own physiological state. Learn to trust your instruments and communicate any potential illusion challenge to a flight mate if available.

Have a crew-briefed plan to counteract common problems like the "duck under." For example, use altitude callouts at night, use two-person scanning techniques (one pilot in and one pilot out), back up VFR approaches by setting navigation aids for an IFR approach. Perhaps most important, stay proficient and know your limitations. Even the best pilots get in over their heads on occasion, but they come out of it alive because they recognize it and have the judgment to ask for assistance or execute a missed approach without hesitation.

Afterword:
Why Be Average?

In the history of man, spanning millions years, those of us who are fortunate enough to fly represent less than 1/100th of 1 percent of all men. Think upon that fact for just a moment. Think of all the thousands of generations of our forefathers who gazed skyward at the birds, wondering what it might be like. Now think of all the men and women who have paved the way for us—men like the Wright brothers, Rickenbacker, Boelke, and countless others. Think of the thousands of pilots who have perished paving the way for us—and now think of how casually we approach this wonderful gift.

This Controlling Pilot Error series is a noble experiment designed to identify, divide, and conquer the last large obstacle in aviation safety—pilot error. Some have scoffed at the very intent and effort, citing maxims such as "to err is human" or that the mishaps are the "cost of doing business." I cannot and will not accept either of these pessimistic points of view—not while the means remain to try another approach.

I believe the vast majority of pilots want to improve, but two problems stand in the way. First, they are not *required* to improve, and second, they may not have a clear idea of *how* to improve. For the past 10 years I have attempted to assist pilots in overcoming these obstacles.

I began this attempt nearly a decade ago when I was tasked by an Air Force four-star general to make an attempt to define *airmanship*. After several years of research into the subject matter—looking through a historical lens at what aviators did *right*—I published *Redefining Airmanship* (McGraw-Hill, 1997). This book defines 10 broad competencies that good aviators have combined into a tapestry of airmanship since the dawn of manned flight. The first book drew a good bit of interest, especially on the topic of discipline. I was asked to probe this phenomenon further and did so, publishing *Flight Discipline* and *Darker Shades of Blue: The Rogue Pilot* (both from McGraw-Hill) in 1998 and 1999, respectively. I was somewhat frustrated. What had begun as an attempt to teach airmanship as a set of broad competencies had ended up focusing on the most foundational block—*flight discipline.*

Interestingly, the feedback I began to receive from various corners of the aviation community was that this approach was having some impact. So it is that we began the effort to identify and attack specific types of pilot error. The jury is still out as to whether or not the Controlling Pilot Error series has been successful in its endeavor.

I have spent all of my adult life in aviation, first as a pilot, then as an instructor, check airman, and now safety and operations manager. Throughout this time I have become frustrated and tired of *organizational* efforts to solve what I perceived as *individual* problems. Since

1997, I have traveled the world preaching the gospel of individual accountability. This was regarded by many as a fool's errand. Andrew O'Connor, a dear friend and fellow warrior in the fight for safe skies, once lamented (in jest), "Sometimes I think we would be more efficient just to beat our heads against a brick wall until we are bloody and senseless. The results are the same and it would take much less time and energy."

I still believe that the individual is the key to solving this last big obstacle in aviation and safety, and that we win the war for accountability and personal responsibility one pilot at a time. However, having beaten this drum consistently for nearly a decade, I have now come full circle back to the realization that organizations do matter, and am now planning to turn my attention to attempting to discover how organizations can best foster individual accountability. So this book represents the end of my research and writing efforts specifically for aviators and I thank the reader for allowing the author these few lines to explain.

Fair skies and favorable winds.

Index

About the Author

Tony Kern is Editor of the Controlling Pilot Error series. He is the Director of Aviation Safety for the U.S. Forest Service and the author of three bestselling McGraw-Hill aviation human factors titles: *Redefining Airmanship*, *Flight Discipline*, and *Rogue Pilot*.